Nurseries: A Design Guide

Architecture can inspire young children; the very shape and form of a daycare centre can not only stimulate their imagination but can help children form strong relationships and help promote development.

This highly illustrated design guide presents all the elements of building design that combine to create the very best environment for young children and the people who work with them, including building materials, multi-functional spaces and design scaled to suit small children. For those involved in capital projects, the book provides a practical introduction to acquiring funds for a new integrated centre or early-years setting and also provides a technical guide to integrating features, such as rooms with many different areas, access to the outdoors and choice of fixtures and fittings.

Mark Dudek is a Research Fellow at the School of Architecture, University of Sheffield and is a practising architect. Mark runs his own London-based architectural practice, *Mark Dudek Architects*.

Nurseries: A Design Guide

Mark Dudek

LONDON AND NEW YORK

First published 2013
by Routledge
2 Park Square, Milton Park, Abingdon, Oxon OX14 4RN

Simultaneously published in the USA and Canada
by Routledge
711 Third Avenue, New York, NY 10017

Routledge is an imprint of the Taylor & Francis Group, an informa business

© 2013 Mark Dudek

The right of Mark Dudek to be identified as author of this work has been asserted by him in accordance with sections 77 and 78 of the Copyright, Designs and Patents Act 1988.

All rights reserved. No part of this book may be reprinted or reproduced or utilized in any form or by any electronic, mechanical, or other means, now known or hereafter invented, including photocopying and recording, or in any information storage or retrieval system, without permission in writing from the publishers.

Trademark notice: Product or corporate names may be trademarks or registered trademarks, and are used only for identification and explanation without intent to infringe.

British Library Cataloguing in Publication Data
A catalogue record for this book is available from the British Library

Library of Congress Cataloging in Publication Data
Dudek, Mark.
Nurseries : a design guide / Mark Dudek.
   pages cm
   Includes bibliographical references and index.
   1. Day care centers–Design and construction. I. Title.
   NA6768.D83 2012
   725'.57–dc23
   2012016256

ISBN: 978-0-7506-6951-1 (pbk)
ISBN: 978-0-08-094092-2 (ebk)

Typeset in Univers
by Wearset Ltd, Boldon, Tyne and Wear

Printed by GraphyCems

For Amy and Grace. My inspiration.

Contents

	Introduction	1
1	Environmental psychology: how to evaluate quality within the learning environment	6
2	The sustainable nursery: a good environment is a natural environment	31
3	The natural child: under a crab apple tree …	79
4	A historical overview: form becomes feeling	110
5	A nursery brief: a machine for learning	143
	Notes	191
	Index	195

Introduction

Raising children is a rewarding but exhausting process. There is no getting away from the transformational nature of being a parent. Many of us fail to be

The issue of scale is fundamental when designing for children. Here the entrance area seat is set at two different levels, 290.5 mm above floor level and 490 mm for adults.

Hodge Hill Children's Centre by Mark Dudek Architects

the parents we want to be, and I suspect that each generation, to a certain extent, repeats the successes and failures of the previous generation. However, today the world is a global and ever more frenetic one, usually dominated by short-term commercial expediency. People are less secure in their jobs, communities are more fragmented. This inevitably impacts on family lives; things feel less certain, more impermanent.

Seen in this context, modern lifestyles can be very stressful for children. This can be significantly relieved by good nursery provision. Simply adapting an existing building, such as a church hall, which is often an expedient approach to low-cost childcare in the United Kingdom, fails to recognize the rights of young children to their own space, and the need to support and reassure parents in every way possible. Equally, it demeans our view of the role of architecture as a power for good in society.

Despite its many shortcomings, it is my view that a coherent system of early-years care and education are the most important political and

Introduction

Similarly, this reception desk is at child height and adult height, giving children a sense of belonging.

This child-height door gives access to the children's play area. Coat pegs are next to it for convenience at child height too. (Both projects designed by Mark Dudek Architects.)

social interventions of the past 15 years in the United Kingdom. Although I emphasize the people involved over and above any building as being the most important and critical factor in this, an environment which enables high-quality care and education to take place is an important aspect of this offer. Many of the 3,500 children's centres that have opened since 1996 are testament to this concern for children and the value of society as a whole. As the social structure of Britain becomes less equitable, so its social problems resemble more those of the post-industrial American cities, with many teenagers excluded from society, and particularly boys lacking male role models. In my view this is a direct result of the ineffectual nurturing during the early years, and the failure to support impoverished and poorly educated parents, itself probably because of second-rate early-years care in their own childhoods.

It is hard to place a price/value on long-term investment in good-quality nursery care and education. There are so many variables at work, and no systematic research has been undertaken that pins down the value of nurseries, let alone the importance of their architectural quality. All we know is that a building is required, and if it works well, then the circumstances of the users will be improved. The historic HighScope project in the United

Introduction

The seat used in these comparative studies, is 29.5 cm high as recommended for 6–8-year-olds (BS5873). Grace who is 4.2 years old, (top sketch 1 and 2), can sit on the seat but not for very long; within five seconds she has repositioned herself supporting her body with feet on the higher intermediate foot rest. By comparision, Amy who 6.2 years old, is comfortable and stays in position 1a for 15 seconds. She can easily adjust her position when asked to look towards the camera, sketch 2b, by simply moving her legs and feet, which are in constant contact with the floor, across.

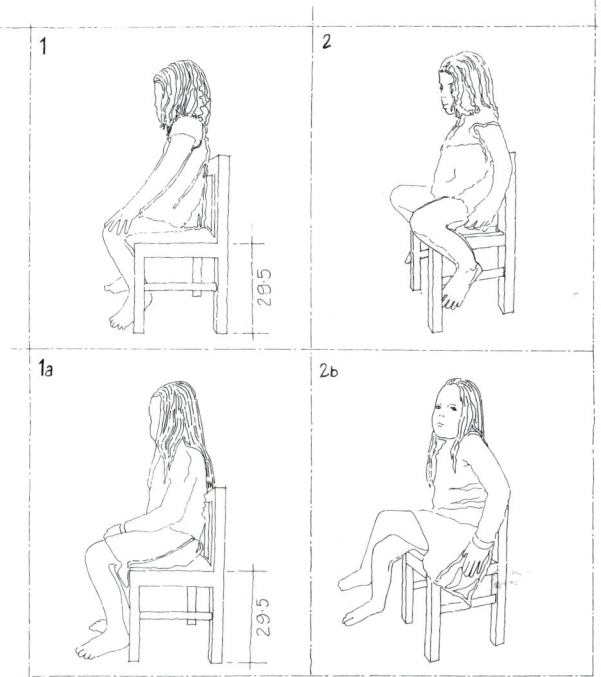

Amy stands at 'her' door at the Yiewsley Children's Centre, designed by Mark Dudek Architects (2010).

States, which tracked two sets of early-years children over a 30-year period (starting in 1964), is widely quoted. A huge amount of help was offered to one sample of disadvantaged families – extra tuition, therapeutic and practical support for mothers, every kind of welfare benefit. Nothing special was done for a comparison sample, and 30 years later the HighScope graduates had done dramatically better. For every dollar spent up front on HighScope, seven dollars were saved further down the road on the costs of police detecting crimes, judicial process, incarceration and those gained from taxes paid and benefits unclaimed.

The conclusion is that as long as the proper resources are provided to help them and offer good-quality early-years care, hundreds of thousands of children will not wreck their own and others' lives. What we hope to do here is explain how the environment not only supports this end, but also show how it can in some circumstances become the most important element in the equation.

A note on methodology

The chapters in this book represent a synthesis of my observations and research over 20 years working in this field. Key recommendations take into account as many as possible of the comments received from interested parties with whom I have debated long and hard during this time. In addition (and most importantly), three other sources have informed my recommendations.

First, research by others. Children's environments research is a growing discipline with key influences cited within the text when used to support the narrative. Second, my own experience of designing and building at least 15 new/refurbished nursery buildings and/or children's gardens for both public and private clients over the past 15 years. Information gathered following occupancy has been especially valuable. Finally, and by no means least valuable, has been the experience of watching my own young children, with half an eye on this publication. Thanks to Amy and Grace.

The recommendations are not cast in stone. Rather, they should be viewed as an evolving theory which aims to prioritize issues in a systematic way. As anyone who has had a hand in developing a new or refurbished nursery building could tell you, compromise is usually the name of the game. The architect, in conjunction with the client/users is there to choreograph competing needs. This publication makes reference to some of the very best international examples of early-years architecture. However, the perspective is very much a British one and focuses on the latest government-funded children's centre initiatives that are currently nearing completion.

There are five chapters, covering a number of linked areas, though each views nurseries and children's culture from a slightly different perspective. Chapter 1 focuses on the discipline of environmental psychology, disseminating some of the key lessons it has learned by close observation of human (and animal) behaviour, to provide helpful conceptual ideas about the nursery environment.

Chapter 2 is a brief summary of the key ideas that go to make up the sustainable nursery, using observations of some examples of nurseries in use to understand what is possible and also to appreciate common shortcomings. Chapter 2 also analyses the nursery school curriculum to explain how this informs architectural spaces.

Chapter 3 asks the question: how do we design for play? This is perhaps the most crucial aspect of nursery design, and is explained with some practical examples of how children relate to the environment in a positive way. There is also analysis of play outside by way of my colleague Susan Herrington's original research at the University of British Columbia.

Chapter 4 explores historical concepts that initiated design for children – starting in the eighteenth century, where early-years care was adopted as a moral crusade by the pioneers, right through to recent developments both in the United Kingdom and in Europe, which hold important lessons for the future.

Finally, Chapter 5 is a briefing document described in terms of the development of a new nursery building in Central Europe. It explains the process we went through to arrive at the final design, and how some of the concepts raised in Chapter 1 have been applied. It also includes detailed technical information which should be useful to anyone embarking on a new nursery building.

At the end of each chapter the reader will find a systematic designer's checklist of ideas and features, distilled from the text.

I would particularly like to thank Canny Ash and Phil Meadowcroft for their inspiring insights during our discussions, Peter Maxwell and Lucia Hutton at CABE for their promotion of 'the cause' and John Allen for his unswerving energy and optimism during my involvement in the Hillingdon Children's Centre projects.

I would also like to thank many other people who have contributed their expertise directly or indirectly over the years, including Eva Lloyd and Alison Clark at the Thomas Coram Research Unit, University of London, and Professor Helen Penn at the University of East London, Professor Cathy Burke at the Faculty of History, University of Cambridge. Rosie Long, Head of Windham Nursery, who has significantly aided my knowledge and understanding of details within the childcare environment (in particular in relation to the sand-pit at Windham).

Recognition goes to the School of Architecture, University of Sheffield, where I am engaged as a part-time Research Fellow. Without their support this publication would not have been possible.

Chapter 1

Environmental psychology

How to evaluate quality within the learning environment

It is now almost 20 years since I made my first visits to see a range of new purpose-built children's centres in Frankfurt, Germany. As an over-excited young architect, what attracted me were the high-end designers commissioned to create the new buildings. Up-and-coming names such as Toyo Ito and Bolles Wilson were commissioned to set the tone for a state-wide initiative aimed at sending out the message that children and families were of the highest priority. The new buildings were to be shiny and new, grand architectural statements, expressions of the architect's ego as much as the city fathers' visionary ideals. While visiting his signature building, one soon to be celebrity architect even told me that it was important to have the building photographed before the users took possession, 'as they would only ruin it'.

If irony was intended in that comment, it was certainly lost on me. The same could be said for those using the new facilities when they finally opened; they often found that the drawing board architecture did not work out for them as intended, the buildings were a little too austere and the children and families for whom they were intended found them cold, clinical places to be. However, I believed that they were a well-funded beginning and like any good building they merely needed time to bed-in. The landscape was immature and the structures set out within the architect's brief did not always reflect the emerging needs of the users. It occurred to me that learning to live with a building was as important as the architect's initial concept. The first five years of any building's life, particularly a childcare environment, is the minimum time frame required for the users to adapt and grow their environment. Yet architects and architectural critics rarely re-assess a building after the first shiny few weeks following hand-over.

During the intervening period of time, I reflected on how these Frankfurt buildings might have bedded down, and revisited some of the early-years examples I had designed myself to check how they were functioning in use. The conclusion I drew was that almost all of the facilities were proving a little too inflexible, perhaps too cellular in plan, with enclosed rooms dedicated to a particular function rather than being oriented to free use and adaptation. I decided that the test of any good children's architecture should be its capacity to develop and evolve together with its users, following a loose fit, long-life concept. The building should not be aloof, like some form of austere beautiful sculpture; rather, it should develop a more personal relationship with its users, becoming a sort of friend and partner, capable of adaptation, change and growth over time.

One of the Frankfurt architects described his building to me as being like a designer boutique or an art gallery. In my view the modern nursery building is more akin to an artist's studio than it is to an art gallery. It is a workshop environment for making and doing, usually messy things, but one that is also calm and reflective in its own right (I will explain the key activity patterns that often dictate the form of a children's building in Chapter 5). It should envelop its users in a warm, reassuring ambience. This reflects the idea that there is a multi-layered poetry to what many might view as a somewhat banal form of civic architecture. The nursery should be like an unfinished story, enabling each child to bring their own fresh response to its narrative verses. Clearly, this implies significant challenges for any designer working in this special area.

Therefore, rather than presenting here only the latest 'state of the art' projects, the aim of this publication is to re-visit a number of those first-generation children's buildings, some of which have now been in use for 10–15 years. Some are even older, and the clues they hold in terms of changes and adaptations implemented by the users over the years are as important as the architect's initial visions. In some instances there was no architect involved at all, rather the building has emerged as a resourceful adaptation out of what would originally have been a building totally unrelated to its final function. Evolving the special culture of a nursery is the talk of all of its users; sometimes the best nurseries, unlike almost any other building type, evolve in peculiar almost anarchic ways.

There is, I believe, no single prototype for the perfect nursery. As we will see, each is a particular response by those involved, relating to a special set of circumstances, some site-specific and some to what might be thought of as people-specific. The basic nursery and its wider community-oriented siblings – such as England's children's centre developments, the Kindertagesstatte in Germany and the Scuola dell'Infanzia in Italy – are so inextricably related to those who run and utilize their services that change and flexibility within the framework of high architectural expectations is the single most important quality indicator. As I believe it is people that make the environment work, their ability to effect the changes to the environment the architect initially gives them is fundamental. This essential understanding lies at the heart of my thinking; it is what makes the contemporary children's environment unique and significant, not just for those who use it, but for society as a whole.

The first key idea I wish to convey is that in an age where novelty for its own sake appears to be one of the supreme cultural values, architecture for childcare is not primarily concerned with the cult of the new, as much contemporary architecture must be by definition. Usually it relates more to

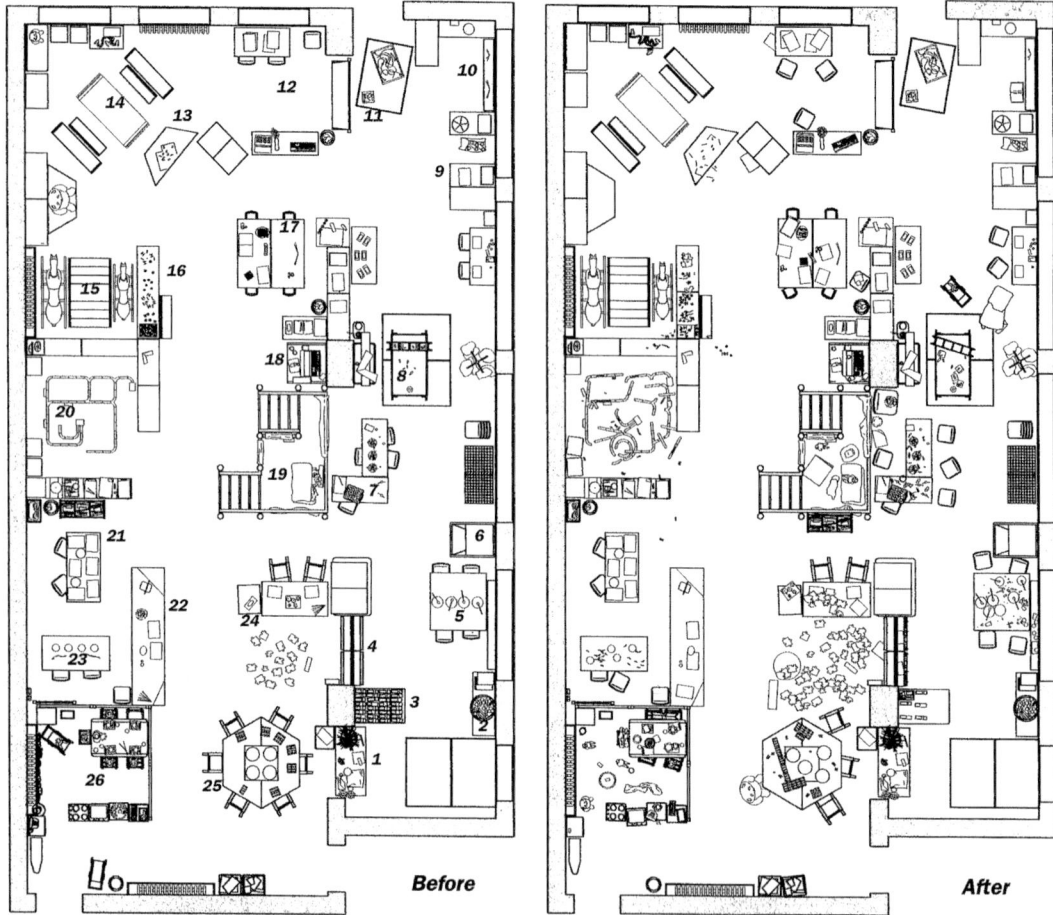

1	Interactive display for Chinese New Year.	
2	Fruit basket – children bring in fruit daily for consumption at break time.	
3	Name table – children recognize and take their name card from the table and put it in a box to show they have arrived.	
4	Easels for free painting – to encourage creativity and fine motor skills.	
5	Malleable activity, making food for birds – science activity, looking at appropriate food for the birds and changes brought about by stirring it and adding fat to make it solidify.	
6	Bugs in the sand – creative play to investigate what has been hidden in the sand.	
7	Making birds for display from carboard tubes, feathers and paper – activity to develop fine motor skills, cutting, sticking and folding.	
8	Water play – based on the story of Mrs Wishy-Washy washing all the animals.	
9	Gluing – an area for independent work using recycled materials to make models, and opportunity to practise fine motor skills, cutting and sticking.	
10	Science area – interactive display of animals in cold climates with books, animals and water, and pipettes to experiment with.	
11	Three-dimensional maze – opportunity to practise fine motor skills moving coloured balls along the maze.	
12	Office area – to encourage independent writing with a supply of writing materials, paper, card and chalk boards.	
13	Small world, farm animals and a house – role play looking at homes for animals.	
14	Book area.	
15	Rocking horses – for imaginative role play.	
16	Poppa beads – to practise fine motor skills making shapes and patterns.
17	Free drawing and emergent writing, based on the book Daisy the Duck.
18	Computer – selected program to complement current topic.
19	Loft – for role play activities, set up with Ten in the Bed big book and animals.
20	Brio train track, to encourage cooperative play.
21	Matching animals activity.
22	Eggs in the nest game – 1:1 counting game.
23	Linking elephants – counting game.
24	Puzzles.
25	Construction activity – H shapes.
26	Home corner – role play area.

Two kinds of change within the nursery environment: a playroom at Windham Nursery, Richmond (designed by Mark Dudek Architects), before and after play. The range of activities indicates a highly proscriptive agenda, partly a response to the researcher's presence perhaps; evaluation by Mark Dudek and Gilian Wardle.

Environmental psychology

Snack time at the Windham Nursery, designed by Mark Dudek Architects.

the capacity that the environment has to grow and develop alongside the evolving patterns of its hosts, especially those of the children themselves, in a tidy and modest form. If a building appears tired and neglected after five years of use, it is a sure sign that it has been designed as a static moment in time, rather than as a vibrant organism with design and funding systems in place that make it capable of flexing to the needs of its users over subsequent years. This shows that design can and perhaps should be a continuing process, with the building growing along with its users, much as a family home will change over the years of the owners' lives there, even if it is in small, gradated stages.

This is not to say that this modest architectural expression cannot produce innovative and even iconic architecture. Far from it, the nursery has scope for the most imaginative architectural invention, which at its best is inspirational. It is simply that architecture for childcare emerges from a different set of influences to most normal building design for adults.

The second key idea that lies at the heart of this publication came to me a few years ago, when I visited my old family home for the first time in almost 40 years. It is, I guess, fairly common to grow up in a small town and then move away as personal horizons change. Yet the environment where we have our most formative experiences, aged 4–10 years of age, is lodged deep somewhere in our psyche. As Sancho Pansa said, 'A man's true home is his childhood.'[1] However, it is often a complete surprise when we re-visit our childhood haunts as adults. At least, it was for me.

The scales and qualities of the backyard where I played as a young child, viewed through my adult eyes were far less vivid than my childhood recollections of essentially the same suburban landscapes. Certainly, something strange had happened to the size of things. In my mind's eye, I remembered one particular environment of my childhood as an undulating mountain range of a landscape full of places to explore, from high up on the ledge beneath the garage to the shelter of the enormous rhubarb 'trees' with the aromas of mint and lavender wafting through from the adjacent herb patch. Now everything seemed ironed out, featureless and rather flat. Structures such as the high gate from which I could climb up onto the garage roof, which dripped icy 'stalactites' during the freezing cold winters, did not seem high anymore. The trees where I constructed my complex treehouse structures were bigger but felt less like the jungle eyrie I had imagined them to be. The postcards I'd pinned lovingly to the timber dwarf walls (to make it seem more homely), which would within a few days become the perfect nesting place for families of earwigs; removing the postcards and squashing the innocent creatures became a perverse pleasure. Now, my adult view could only discern a slightly overgrown suburban willow hedge. My magical childhood perspectives, admittedly tinged with a large dollop of sentimentality, had inevitably been transformed into a pragmatic adult view. It was a disappointment to say the least.

A second type of change relating more directly to the built environment. Windham Nursery, by Mark Dudek Architects, before and four years after. Initially the environment outside is bare and uninteresting, while four years later aromatic planting boxes raise its sensory quality, and a rudimentary plastic canopy extends the field of learning.

These memories are, of course, subjective observations tinged with the sentimentality of a happy childhood lived out 40 years ago. But are there any less subjective and more scientific assessments of the need for good childhood environments which might help to define what this might be, away from the world of the architectural criticism and subjective observation? Looking towards the discipline of environmental psychology, there is a huge and ever-expanding literature of research on environment–behaviour transactions. One must ask why so little mention is made of it within the architecture and planning professions; surely the most likely potential users of the findings in shaping future designs. Do they not know about it, or do they simply ignore it, given all other demands upon their attention?

The architect Thomas Fisher, in an article with the splendidly ambiguous title 'Architects behaving badly', suggests two apparently opposite reasons.[2] First, that all the findings of environmental psychology are so obvious as to be trivial ('Of course, people like nice views from their windows; enjoy greenery; and like places where they feel safe, while being able to see out'). Second, that the research findings are presented via inaccessible academic journals, written in abstruse language, cluttered with literature precedents, formal hypotheses, with an emphasis on data tables, data analyses and statistics, and with conclusions all hedged about with ifs and buts.

To the first point Fisher replies (to his own rhetorical questions): if these good suggestions are so obvious, then why have we not seen their products routinely present in all recent design from the professions? The second point poses the greater challenge: findings must be made more accessible and that awareness services must bring them to the attention of the professions. This is now happening in the work of UK organizations such as CABE and English Heritage, and new, easily accessible indexed summaries of research should make it as routinely accessible as all other information sources (for example, on products, materials, legal matters, etc.) already do for the designer and architect.

My friend and former colleague Emeritus Professor Christopher Spencer, who is based at the Department of Psychology, University of Sheffield, provides a very erudite explanation not just of the world of the nursery, but also the wider urban environments of the street and the park, where children spend much of their lives. Here I am using an article he produced specifically for this publication in its entirety. He starts by describing his early research into animal behaviour:

Once upon a time, in a Malaysian jungle, I was a field worker studying the largest of the lesser apes, the siamang (which looks like a larger, black gibbon); and as a social psychologist, I was most interested to observe the stable family group and its interactions. Typically, there would be a mated-for-life pair of adults, a confident juvenile, and a recent infant; a family holding a large area of jungle as its exclusive territory; and spending its time foraging for fruits and leaves high in the canopy.

Spending one's days beneath such a family (in our study site reserve, the animals were long-habituated to respectful observers), one quickly became aware of the infant working through the tasks of early childhood, heightening one's awareness of what faces the human young.

Environmental psychology

Leaving the comforting immediacy of the parent (a small irrelevant aside: it is the male siamang who typically specializes in most of the child-care), the infant has to work out its physical capacities and the effects of its movement on the perceived properties of the local world; although still being suckled (still the female role!), it is testing the properties of things encountered, what tastes good, what is to be avoided, what you can expect to be where (water trapped in tree ferns; stinging ants in a marching column). And as the infant grows more confident and adept at mobility, there come issues

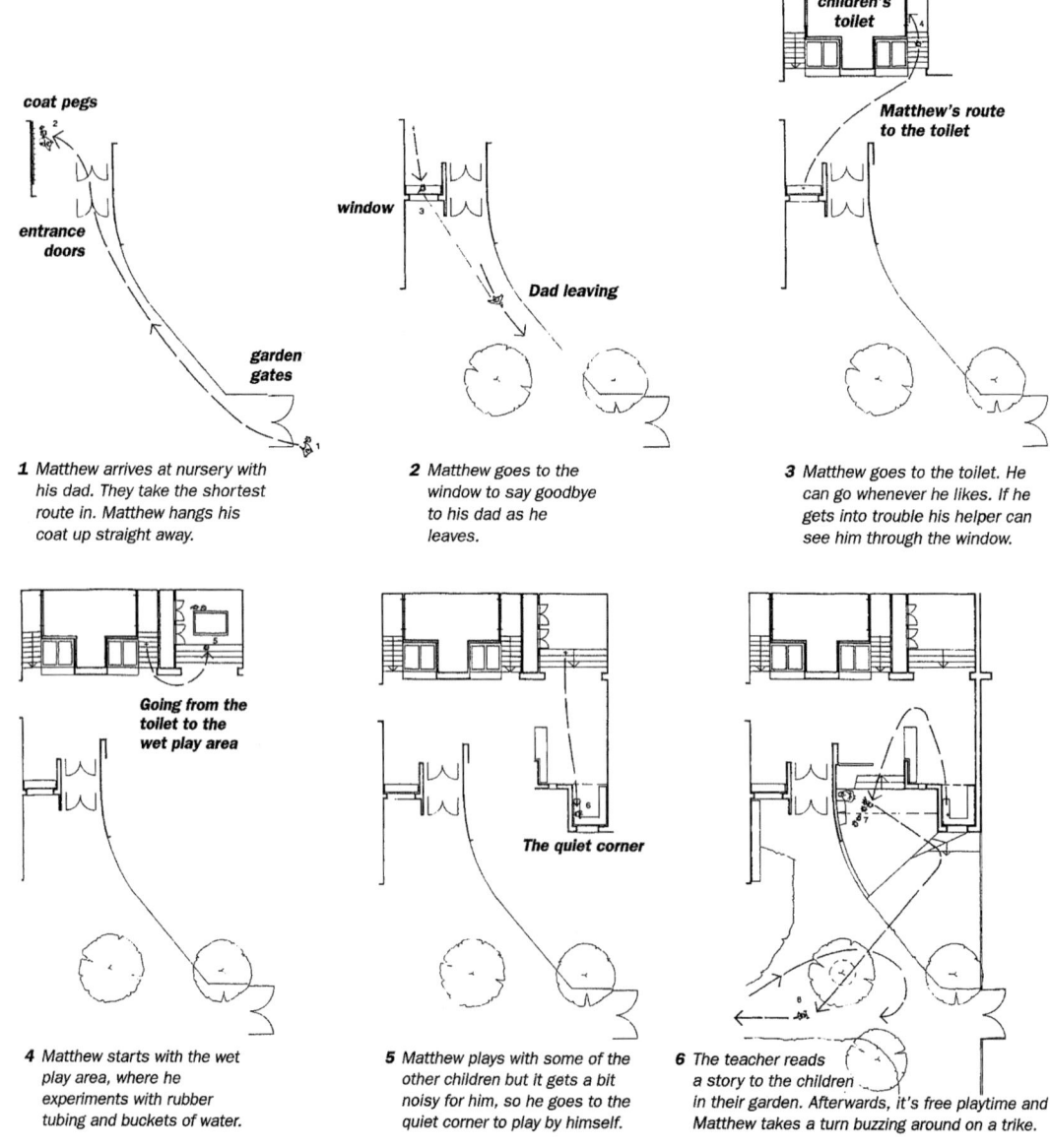

1 Matthew arrives at nursery with his dad. They take the shortest route in. Matthew hangs his coat up straight away.

2 Matthew goes to the window to say goodbye to his dad as he leaves.

3 Matthew goes to the toilet. He can go whenever he likes. If he gets into trouble his helper can see him through the window.

4 Matthew starts with the wet play area, where he experiments with rubber tubing and buckets of water.

5 Matthew plays with some of the other children but it gets a bit noisy for him, so he goes to the quiet corner to play by himself.

6 The teacher reads a story to the children in their garden. Afterwards, it's free playtime and Matthew takes a turn buzzing around on a trike.

There are six stages of movement within the nursery. This schematic shows four-year-old Matthew's fluid route around his nursery (drawn by Simon Price). This is natural for children (as opposed to most adults); they are almost constantly moving from one activity to the next and so clear pathways within the nursery are critical.

Environmental psychology

The critical interaction between movement and perception or learning by copying.

of navigation and location, away from the familiar and comforting, and rapid returning if danger threatens.

Growing up in a nuclear family, isolated from other groups, and having no age peers wherewith to interact, the siamang infant might seem to inhabit quite a different world from the human child in nursery school: no chance of the full pedagogic experience here!

Yet with regard to the importance of space in early development, we can draw clear parallels: the interaction between movement and perception; the learning of the properties of things and their location; learning the typicalities of location and what has been described as the grammar of space; exploration while being aware of refuge; basic wayfinding skills. Where are the resources? Indeed, how can I find my way around the place? Where is safety and comfort when I need it? Where can I go wild and test my physical prowess? Where can I be quiet when I want to? (In siamang terms, how can I avoid the juvenile?)

If siamang and human infants have in common a need to learn the dimensions of space – testing ones physical and cognitive capacities against its challenges, and wayfinding in socio-physical space – then one can also see that for the human infant the nursery school poses still further tasks, with its larger numbers of individuals to negotiate and greater range of resources both to use and to dispute.

Put like this, life in the canopy seems somewhat less daunting than that in the nursery school!

As designers, how can we respond to the needs and challenges of infancy in company but away (most probably) from the most familiar caregivers? What are the physical and social design features that will support the child? Individual needs and the social purpose of the nursery may need to be in delicate balance: exuberance and quiet; collaborative and social activities balanced against those occasions for focused and individual play.

Environmental psychology

Whether talking about humans or lesser apes, one realizes that the youngest infants and the slightly older individuals make different demands on space, so in the mixed-age nursery, one realizes that design will have to take into account interactions between younger and older infants. How is this to be achieved in design terms?

Environmental psychology has as its brief the elucidation of the transactions between people and places ('transactions' because the relationship

The playground at Discovering Kids Nursery in Loup, Northern Ireland (designed by Mark Dudek Architects), a place where I can go wild and try out my physical prowess. However, where can I be quiet when I want to? The answer is: on the raised play-deck running along one side of the playground. It is important to make provision for both types of space, both outside and inside the nursery (see the plan on page 64).

Environmental psychology

The nursery can be a place for experience and imagination, of juxtaposition and novel experience.

works both ways: as Winston Churchill said, we shape our places and they shape us). Some of environmental psychology's concepts may be of use here; it describes such ideas as place attachment, prospect-refuge, cognitive maps as used in wayfinding and resource location, sociofugal and sociopetal spaces, the way the layout of some places keeps people apart, while others bring them together. It researches the evidence for such widely held assumptions that, for instance, the natural world is good for one, is restorative (whether as directly experienced, viewed from a window or even as just represented in a picture). Always seeking evidence for its claims, environmental psychology has given an empirical base to the aesthetics of place, showing the dimensions and characteristics which people find most pleasing. This might sound remote from the world of the young child – but not so! The designer needs to know what features can effect stimulus and which can calm, can interest and involve, can welcome the newcomer and maintain attachment in the old hand.

Is there a place equivalent of pedagogy? Does early experience of rich, varied and interesting places have an effect which sensitizes and lasts into later life? In his writings, Ruskin always spoke of the enrichment that came from exposure to beautiful things; and his collections of diverse pleasing objects and images was the centre of his popular educational enrichments for the English working class, juxtaposing the peacock's feather, the lapis lazuli and the image of Venetian decorative stonework.

Children gather around the pond at the Portman Early Childhood Centre, Westminster, London. New garden and decking designed by Mark Dudek Architects (2003).

So similarly, the nursery can be a place for experience and imagination, of juxtaposition and novel experience. And those experiences can be of the natural world as well as the man-made, decorative; seeing the living plant and animal world can give more than the restorative effect already mentioned, especially if the connection with the child involves the child caring for and having some responsibility for the nursery nature. (Recall the evidence that early experience of caring leaves its trace in adult character.)

A related and almost political point which one might consider here, especially in the case of infants and young children from 'deprived' or 'underprivileged' backgrounds: the nursery school might represent an island of enriching experience in the child's life.

The concepts and supporting evidence that environmental psychology can offer our understanding of young children and space

Since the 1970s there has grown a large empirical environmental psychology literature that is of relevance here. From this, I will offer some examples of concepts from the field, drawing on evidence, that:

1   'green is good for you';
2   aesthetics and attractiveness are predictable and rule-following;

Grace clasps the plastic rolling-pin in one hand while feeling the grainy birch ply of the table-top with the other hand. These intuitive sensory experiences are part of her environmental awareness development and provide critical 'affordances'.

As soon as she can walk, Amy heads for every uneven or challenging surface she can find during her assisted walk to the shops. This pavement edge provides her with 'affordances' of objects and places which help to develop her natural balance and coordination.

Environmental psychology

1.4-year-old Grace carefully negotiates the rapids at the Princess Diana Memorial Fountain. The feature provides contrasting experiences of cold feet on rippled, textured granite underwater, a challenging experience for young legs. Further down is a much safer walking route, and the same surface in its dry condition warmed in the afternoon sun. Aesthetics and attractiveness are predictable and rule-following.

Efficient way-finding relies on good cognitive maps, the experience of a novel feature available for sensory exploration within the general environment can significantly aid this knowledge for young children. The girls generally stop at the two telephone kiosks on the way to the park and spend 2–3 minutes speaking into the handsets and pressing the buttons. Similar orientation points can be provided within the nursery.

3 efficient wayfinding relies on good cognitive maps;
4 children in their everyday life are good at discovering the affordances of objects and places, going well beyond what their designers had intended;
5 studying human ecology may predict a child's behaviour better than any knowledge of the child's personality or upbringing.

Finally, now that all this good and relevant research is there in the published literature, we can ask why the design professionals (architects, landscapers, interior designers) seem to be ignoring it. Is it that social scientists keep on coming up with statements of the obvious? Or, rather, that their literature is so complicated and jargon-laden … and that, anyway, designers think in pictures rather than words?

1 'Green is good for you'

This is a proposition that most will concur with, without stopping to consider what evidence there is. Yet we can now do better than this. When environmental psychology was young, there was little systematic evidence to call upon; and what did exist was of unclear validity. Supposing an early survey compared the health and well-being of young children from leafy suburbs with that of those living in inner cities: almost certainly the former would score more positively than the latter.

But could one attribute this to the more 'natural' character of the suburbs? What else might differ between the two settings? Average size of the dwellings; size of the rooms therein; consequent availability of privacy and 'own space'; average family income; ditto educational level; parental health; air quality; particulate pollution; noise from traffic and from neighbours; perceived and actual levels of crime and incivilities; 'stranger-danger'?

So one would seek less unambiguous evidence for the proposition. For much of psychology, the answer would be to bring one's phenomenon for study into the laboratory, and control all other variables but the one under study; but obviously the reality of a child's environment isn't susceptible to capture in the lab.

As a researcher, one might dream: if there were only real-life settings where all the variables happened to be held constant but for this key one ... but occasionally there *are* just such 'natural experiments', and early environmental psychology began to find them.

Famously, Roger Ulrich found one when looking at well-being and recovery rates in (adult) hospital patients: people with similar demographics and medical conditions, allocated to wards with either a view over parkland or of other hospital buildings. Not only the patients' self-reports of their well-being and contentment, but also the more objective medical indicators of recovery went in the expected direction: green (views) *were* good for them.

More recently, in a whole series of carefully described settings, Frances Kuo has demonstrated the wide range of positive effects that exposure to a 'green' environment can have for children: these include the calming effects of green surroundings on children with ADHD and ADD, and on rates of aggressiveness as well as on ambient crime in the community (closely and probably causally related to improved community cohesion) as it might affect the quality of children's lives.[3]

Many of the Kuo findings come from 'natural experiments': where all but our key 'green' variable is held constant. For example, in one public housing project in the south of Chicago, the monies for landscaping ran out halfway across the project; this means we have the same housing stock and similar tenants, but half the project with good communal gardens and half with just open grasslands around them. Using standard measures for assessing the children's well-being and behaviour, any differences could be reasonably attributed to this difference in ambient greenery. Visit the project, and you would see more neighbourliness in one compared to the other; see fewer instances of graffiti; hear fewer reports of crime and more positive observations by those who live there. All these are indicators of well-being.

2 Aesthetics and attractiveness are predictable and rule-following

Could we predict what young children will find attractive and pleasing in a new nursery? Once upon a time, aesthetics were held to be such a matter of individual response that it would be impossible to have a science of aesthetic preferences, but environmental psychology now maintains that it can be done; and that we can meaningfully ask what designs would be pleasing, aesthetically, to the children.

We still need careful research with young children here; all the work so far published is with adults, but it is promising. It shows that there is a predictability about the features which people find pleasing. Thus, for example, the presence of water (such as a stream or a lake) in a landscape view will increase the likelihood of the view being reported as pleasing compared with a similar scene without water. A typical experiment might offer a large array of photographs – some with a stream and others with the stream photoshopped out – and ask people to sort them according to preference.

Environmental psychology

Complexity is good. The Cherry Lane Children's Centre by Mark Dudek Architects. Its in-built complexity, which is also reflected on the external façades, offers the gift of enhanced play to its young users. An optimal level of complexity equals mystery and therefore an awakening of curiosity. However, complexity must not go too far to the point of disorientation.

One would make sure that each person would only see a particular scene once, either with or without water, so that suspicions are not aroused.

Many other features have now been systematically investigated: complexity, 'mystery' and novelty being examples. A complex scene invites exploration more than one which is simple and immediately readable. This is why, says Kevin Lynch[4], we as tourists are tempted to investigate the half-hidden features of an Italian hill town. Equally, a playroom with corners and hidden places will hold more fascination for the young child than a straight-forward and easily parsable one. Stephen Kaplan has proposed several rules for predicting which scenes and settings will attract and restore us: these include this feature of fascination; an optimal level of complexity (too much will leave us bewildered and lost); a level of novelty (he talks of 'being away' physically or conceptually from the typical setting); and a compatibility with the person's inclinations (see also 'synomorphy', below).[5]

At the scale of an individual room, it is similarly possible to discern the factors that predict people's aesthetic response; indeed, some of the earliest published studies in the environmental psychology literature were those by Canter and Woods.[6] They carefully disentangled the factors which make us respond favourably to the geometry of a domestic room. In their studies the stimuli shown to participants were a series of line drawings that systematically varied such room features as ceiling height, slope, fenestration and so on.

These were simple, reduced stimuli for research purposes. But later research indicates that if it is possible to engage several of the senses, the effect will be greater: multisensory arrays with a variety of colours, textures, lighting effects, sounds and perhaps aromas (again, as with complexity, one can over-do it!).

Of course, there will be individual differences in aesthetic preferences, but from these and similar studies one can begin to discern clear communalities in response, sufficient to make them useful guides for the designer. We need the same kind of research on young children's aesthetic preferences for shape, colour, texture, layout and so on. I predict that we will find that they will value features which signal comfort, safety, interest, opportunities for exploration, as well as a layout with different phases, with some open areas together and some corners in which to be quiet.

3 Efficient wayfinding relies on good cognitive maps

Imagine the young child entering a new setting: exploration of the layout of the place starts immediately, visually and then by walking around, examining features and relating them spatially. What is going on here is the rapid development of a cognitive map of the new area: which must be a primary survival skill for any organism (think how a domesticated animal explores a new home).

Where am I? Where are the resources, the dangers, the safe places? How can I move efficiently and reliably from point to point within the space? Adults may help, but from the first, the infant is an active learner, exploring, integrating, predicting, experimenting. What has been called the 'legibility' of the setting will support this process of learning and developing the cognitive map. The first person to write about legibility and link it to both memorability and aesthetic value was Kevin Lynch, the great theorist of town planning. But what he was discerning at the level of cityscapes can equally be applied to the scale of a building or a room in a nursery school.

It is easy for adults to underestimate the cognitive mapping capacities of the young child, so used are we to acting in a protective and undemanding way when we travel with them, allowing them little opportunity to demonstrate how well integrated their spatial knowledge has become into an efficient map.

Environmental psychology

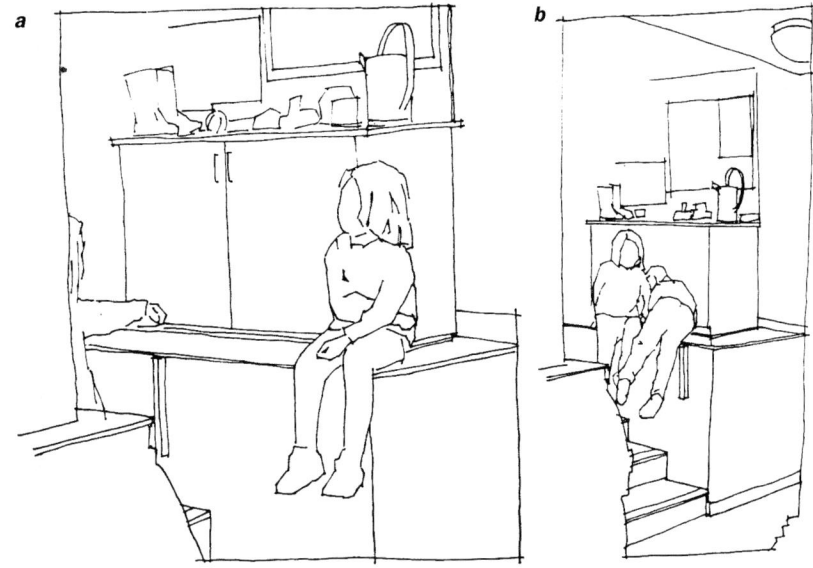

Children using a ledge in a Sheffield day-care facility (from observational studies by Simon Pryce). (a) A child sitting on the ledge is spotted by her friend. (b) The friend climbs along the ledge to take up the position now being vacated by the first child. An example of a non-designed feature that enables children to inhabit the space in their own way

Yet, as Zhra Darvizeh and I were able to show in a series of novel neighbourhood walks with three-year-olds, when we feigned ignorance of the route to be retraced on the way back to the start, the child was able to recall key junctions and decision points. We even pushed them further, and asked them back at base to use toy houses and roads to make a physical map of the route travelled; they weren't very accurate – unsurprisingly – but the task made sense to them, and evoked clear memories of the walk.

Indeed, the geographer Jim Blaut maintained that we should consider the very act of mapping, translating a good cognitive map into an actual physical map to communicate spatial information to others, as a cultural universal, so important a survival skill is it for people living in social groups.[7] In testing this proposition, we have found that children as young as three-and-a-half can follow quite complex routes using a map that shows the route across a playground-maze.

In our nursery schools we can help exercise this capacity by challenging the children to locate and relate features of the space, using the well-designed complexity of the setting in our pedagogy.

4 Children in their everyday life are good at discovering the affordances of objects and places, going well beyond what their designers had intended

There is now a growing literature on what young children actually do all day; and this indicates how ingenious they can be in discovering the possibilities of objects and places that even their designers had not realized (how many things can *you* do with a cardboard box?).

In the early days of developmental psychology, children's everyday behaviour was little studied, the researchers preferring the controllable setting of the laboratory (often strange and sterile for the child, who was left trying to make human sense of the tasks the experimenter set them). Indeed, in a book on children's environments we published in 1989[8], I managed to sneak in a picture of our family cat out in the garden, to make the point that at that time we had better systematic studies of what domestic cats did all day than we did of young children's daily life. Things have substantially improved since then!

We now have careful, extensive field studies of children's free activities by researchers from all continents: and a theme which runs throughout is this ingeniousness, whether to use 'spaces left over after planning' for their play; or whether to transform objects (that cardboard box) into any manner of imaginative play objects. Play experts tell us that the worst toys are those highly designed ones which are not open to re-interpretation.

'Affordances' is the term that has been used to describe our ability to see what an object can offer or afford us: maybe going well beyond the first, intended use. (That box isn't just a discarded container – it can be, successively, part of a den; a fire engine; a spaceship; a climbing obstacle; a weapon....)

Characteristic of the pattern of children's play is the free flow from activity to activity, transforming as it goes. Robin Moore, accompanying young people through 'childhood's domain', is struck by the level of discrimination and inventiveness in using resources: 'Each child wove a pattern of personal playtraces through the neighbourhood, laced together with the traces of other known and unknown players.'[9]

Watching children at play in a nursery, one sees this same complexity. Younger children tend to play independently of each other, moving from project to project. Cooperative play emerges later. Robin Moore says[10]: 'Metaphorically, the result looks like a cloth of varied solids and voids, arranged in an irregular geometric pattern.'

There are clear implications for the design and running of a nursery, to foster and afford this fluidity and ingenuity of play and exploration; and to realize that much of the time, young children will not spend their time in long periods of focused activity. This is normal, expected and positive. (Remember the young siamang learning about the world via play, moving from activity to activity.)

5 Studying human ecology may predict a child's behaviour better than any knowledge of the child's personality or upbringing

The most important precursor to environmental psychology was ecological psychology, a subdiscipline started in the 1950s by dissident developmental psychologists who were aware that behaviour in the laboratory was so atypical as to be near worthless studying. Instead, they set out to do systematic studies of the everyday settings of childhood. Roger Barker and colleagues had asked what children do all day, in both the formal settings of school and church, and outside them.[11] They studied the whole young population of Midwest (Kansas) and Yoredale (Yorkshire), two small towns, over many years. They quickly saw the repeated regularities that more casual observers would miss in their brief encounters.

Typically, psychology has stressed the individual determinants of behaviour, maybe seeing it as the product of stable personality traits, or emphasizing the role of parental styles, or particular and exceptional circumstances. But seeing successive generations of small-town children go through the same routines convinced the ecological psychologists that the determinants of any particular behaviour you studied were the set features of both the local physical and the social world: the furniture of the schoolroom; the layout of the town; the expectations of the teacher; the indulgences of the sweetshop owner, etc. When in maths class, children, year in and year out, acted 'maths class'; released from school, they acted playground, drug store, homecoming, bedtime. Generations of them!

There would be seasonal variations in the pattern of playground games, of course; and of course, there would be more and less boisterous and competent members of the age-cohort. No-one denies that time and individual differences exist, but the realization of the quiet, almost background determinism of the socio-physical environment must have come as a surprise to many psychologists (anthropologists trained to look for just such a patterning of behaviour would have been surprised at their surprise!).

So, in our nursery school, we should become aware of this soft determinism of both physical and social features of the behaviour setting; and be aware of how we can use this to facilitate the good workings of the nursery. Clearly, there are some physical settings which are more conducive and supportive of the intended atmosphere; a draughty church hall may present more challenges than a purpose-designed setting. Ecological psychologists talked of a synomorphy, a goodness-of-fit between activity and place; they also studied the effect of size of institution upon the behaviours within: for example, a big school experience versus a small school, both ostensibly with the same remit and intention, but with predictably different levels of involvement for the individual members.

Spencer does not, of course, explain in practical terms how these findings may be translated into how we might form genuinely high-quality childhood environments. However, the lessons to any sensitive architect regarding the nuances and subtle complexities of the true early-years learning environment are clear. In particular, Spencer's comments on what spatial features enable children to play, or the 'affordances' which promote learning within very young children are cogent and will be dealt with in more detail later. However, I believe that the discipline of environmental psychology is incredibly helpful when dealing with the conceptual development of any space. Now I wish to introduce another related concept, the notion of the 'child within'.

The child within

As a child I did not attend a nursery. My memories of a childhood made vivid by my own capacity for imaginative play within a found landscape, sometimes shared with the childhood friends who played with me, sometimes alone, but always open and accessible, touches upon the essential quality any children's environment must have – it must support rather than hinder the scope for play. The transformation that most people experience when they progress (if that is the correct idea) from childhood to adulthood and all points in-between, is dependent on a full and active childhood, if the child is to grow into the adult. The inner child is always there to support that adult life.

The concept of 'the child within' has been part of our world culture for at least 2,000 years according to Charles L. Whitfield: 'Carl Jung called it the "Divine Child" and Emmett Fox called it the "Wonder Child". Psychotherapists Alice Miller and Donald Winnicot refer to it as the "true self". Rokelle Lerner and others in the field of chemical dependence call it the "inner child".'[12]

It may seem strange to the casual reader to be referring to someone who is in a 'field of chemical dependence' when talking about a book on buildings and architecture, let alone the even more esoteric notion of the 'child within'. However, it is important for me to stress how broad-ranging my frame of reference is, often going beyond the field of architecture and building. My recent feedback from the Head of Children's Services at a London Local Authority, who criticized my design as being difficult to fit furniture into because of the curved walls emphasizes this lack of understanding generally and a shallowness of ambition for our children.

This engagement with other disciplines (in addition to architecture) has broadened because of the variety of people I have encountered and the often unique ways in which they engage with their environment. My terms of reference for this study are essentially the engagement of the human mind to conceptions of childhood, a wide-ranging frame of reference which takes me to places that may at times seem wholly irrelevant to the task of designing a building for childcare. Whenever a childcare professional explains to me that a particular space or feature within the environment is valued or special, I have tried to search out the deeper meaning rather than go for some sort of easy proto-functionalist explanation. This reflects my view that the world is a rich and complex place, it has only been made to be flat, boring and even for lazy, adult-centred reasons. I seek to divulge the child-centredness of things.

Today it can be argued that the design of buildings for education has never been more orthodox and deterministic. It is determined by tight budgets and standard 'area guidelines' which are usually laid down by central government requirements, which leaves bureaucrats and accountants in charge of this most important area. Architects, while well intentioned, are generally confused as to the most appropriate strategies for good design. Do they listen to adults or to children? Because they have so little opportunity to consult with the juvenile end users, they tend to design to relatively simplistic generic standards. Build it quickly, build it simply and build it cheaply is the general approach (with some notable exceptions), with a bit of sustainability bolted on at the end. There is very little considered localism. Spencer's rich ideas about complexity and variety are largely absent. It is very hard to get excited about new nurseries designed in this way.

Most books about school buildings are produced from a quasi-technical stance. Does it help to deliver the educational curriculum (the right size and number of classrooms, for instance)? Is it safe and secure? Does it help with crowd control? Is it cheap to construct and maintain? What are the

Environmental psychology

The Children's Centre at Yiewsley, designed by Mark Dudek Architects. The clients wanted the entrance area to be part of an open kitchen/café space, which dissolves the institutional feel when entering the building, making it warm, homely and welcoming. Despite opposition from the Local Authority, the user group were determined to retain this important feature during design development, attending every site meeting and never allowing the open kitchen idea to be lost on the basis of spurious health and safety issues.

The mini garden and main entrance beyond.

Refurbished community hall

Entrance and open kitchen

Existing church and church hall

Office and administration

Main entrance

New children's centre

Garden gate from local DIY store

outcomes relating to the initial capital expenditure? In short, it is like a shopping list of lowest common denominator, centred on control, security and constraint. This approach hardly addresses the early-years environment as a living, evolving, holistic community where education is promoted as something to excite and stimulate its children rather than restrain and bore them.

I will argue that good nursery building usually happens when mavericks (rather than bureaucrats) get involved. These are individuals or groups who can carry their user clients with them, designing to a strong ethical or ideological standard dealing in both hard and soft sustainability processes which are essentially child-centred rather than adult-centred. That is why the people involved in the development are as important as the end result. We will explain how the community relates to their environment as well as describing the building and how it supports and reinforces the child within.

Whitfield goes on to say that the 'child within' is the part in each of us which is our true self, it is what we really are, it gives us energy, creativity and fulfilment in our everyday lives as adults. Denial of the child within when we are children, an effect caused in families where there is little nurturing, where chronic physical or mental illness creates a rigid or cold environment, thus denying the child his or her right to spontaneous play, creates a co-dependent adult. If the child has no escape from this type of home environment it may lead to chronic anxiety, fear, confusion, emptiness and unhappiness in adulthood. An extreme view may be and is by no means a blanket truth (for every childhood has certain shortcomings), yet living in this incarcerated way is so obviously inhumane. This cold, unloved parent will merely pass on their own neurosis to their children.

Where my childhood escape was to what was perceived as the relatively unsupervised outdoors, which during the 1960s seemed safe, it is right to ask the question: where do children escape to today if there is no garden, or if the streets surrounding the child's home are made out-of-bounds by adult behaviour and adult perceptions? Bertrand Russell said that nursery is a necessary escape for certain types of children to find their own sense of spatial awareness:

> freedom of movement, freedom of noise, and freedom of companionship ... the right sort of nursery school will only have so much instruction as is necessary to keep the children amused. So far from straining children it should afford them relief from the supervision and interference which are almost unavoidable in small homes.... Children deprived of all these needs until the age of 6 are likely to be sickly, unenterprising and nervous.[13]

It is the social interaction of healthy childhood play which supports a grown-up existence of self-sufficient responsibility. Elsewhere we discuss the nature of play and this emphasizes the social interactions with other children, which today only happens in nursery. I would argue that ultimately it is this rich, sharing experience that enables us to become successful parents in our own right. A definition of the evolution of man is from a selfish, self-centred individual (aged 2–3 years, when everything is mine, mine, mine), to a fully rounded person relaxed with the social needs of the community within which he lives (with a healthy communal view that we are all somehow in it together). Obviously the latter view is one which resonates with the social pedagogy which lies at the heart of the best early-years care. This is what the nursery should ideally provide and the environment is the essential partner in this structure of care and support, which is particularly important for children who grow up in troubled family environments. It is even more important now that children's worlds have seemingly become so enclosed and restricted by the fear of crime and the domination of their environment by motor traffic.

Summary

There is a final thought to add to these somewhat cerebral musings as I re-read my manuscript for the umpteenth time. In an ideal world it would be great to expect the same funding levels for a new children's centre or nursery building as I would expect to receive if designing a commercial building such as an office, a restaurant interior or an art gallery (although as the credit crunch has almost terminated the development of any of these building types for the time being, it is perhaps the past tense in which I should be speaking). Unfortunately, architects in this sector must get used to the tight funding regimes which dictate construction budgets for these largely state-funded facilities. Inevitably, certain aspects of the quality agenda will suffer as a result. However, there are interesting opportunities for certain types of architects who are stimulated by the prospect of using new materials and getting involved with self-build community-oriented projects to achieve interesting buildings for children at an economic budget, in all its various forms.

It is also important for me to emphasize here that some of the best facilities I have visited are those which have been adapted over the years on very tight budgets. For example, the Burley Children's Centre, which is discussed in Chapter 2, is a particular example of a building which suits its purpose without even having originally been designed as a childcare facility (it was originally conceived as a school). As much of its environmental quality comes from features added by the users themselves, such as the DIY picket fences breaking up the open-plan interiors or soft corners conceived and decorated by the staff with left-over materials such as sea shells and plastic bottles, it is important to reiterate that the quality of the environment is inextricably linked to the imaginative engagement of the staff operating within it; its success resides in the staff who are able to think and act for themselves in a similar creative way to the children themselves.

Hard though it is for me to say this as a practising architect, high-end design does not necessarily make for a successful centre; rather, it is the chance encounters facilitated by an environment which is in itself childish, transcending more prosaic adult priorities such as security and visual transparency, offering instead something that children themselves can play with and in. Children are never happier than when they are using the environment in ways that adults had not intended. What I refer to as the 'soft qualities' rather than the 'hard qualities' (health and safety priorities such as supervision and the ease of cleaning and maintenance, for example, which usually works

Environmental psychology

against more child-centred qualities), are the aspects of environmental quality which are of the highest priority. The key to a successful centre is one which at least feels somewhat removed from a centralized stifling bureaucracy. Rather, it is an environment which is, to all intents and purposes, open to manipulation and change by the users themselves.

That is why broadly agreed definitions, which transcend professional disciplines of what constitutes 'poor quality' and 'good quality', are hard to pin down. How does one value an environment which facilitates great play, yet is difficult for adults to supervise? How do you define what constitutes good quality between the context of a daycare centre in a wealthy suburb of Paris against that of a community crèche in the drug-dominated town of Tijuana? However, one of my aims here is to also establish clear benchmarks for what constitutes good quality, albeit viewed from a range of alternative perspectives.

Professor Gary Moore has developed an evaluation system as part of his ongoing research into early-years environments at the University of Sydney. What he calls the Children's *Physical* Environment Rating Scale for assessing the quality of early childhood care facilities comprises four major sections and 14 subsections. The first is *planning*, which covers centre size and modules or home base disposition. The second category is *overall building* and deals with image and scale, circulation, common core of shared facilities, indoor environmental quality and safety and security. Third is *indoor activity spaces*, covering modified open-plan spaces, home bases, quiet activity areas, physical activity areas and messy activity areas. Finally, *outdoor spaces* covers play yards, location and site issues.[14]

This picture of children climbing in the wrong direction in a city park, Boston – sliding up rather than sliding down – was not staged. Merely a typical example of children using the environment in ways adults had not intended.

Each of these major sections is given an overall rating scale which is intended to provide guidance as to the effects on young children of the physical environment. Moore states that although previous research has established a positive relationship between certain physical environment characteristics such as size, density, definition of activity spaces and childhood development, these rating systems focus on the social environment. In other words, although value is applied to these basic characteristics of building, actual architectural quality, what Moore refers to obliquely as the 'image' is not counted. Indeed, architectural flair is often viewed as a waste of money, something for more sophisticated people to appreciate than three-year-old children.

The so-called 'Sinking Boat' Kindergarten by Behnisch and Partner has a very strong architectural image in line with the original brief, which was to place Stuttgart and high-quality childcare on the national agenda in order to attract families to the city during the 1980s. In practice this type of high-end architecture is, for cost reasons, unusual for childcare buildings and not always positive in use.

CHECKLIST 1: KEY CONCEPTS WHEN SETTING-OUT YOUR DESIGN BRIEF

1. Concentrate on natural, soft materials, the controlled use of colour and what I am calling legible complexity. Therefore, some curves are better than all straights, particularly in corridors, as shown in the Cherry Lane Children's Centre. Children soak up social interactions when they are moving between rooms. In a similar way, a more serrated activity area rather than a square box works better. Note the child-height windows and door at the Cornerstone Centre.

2. Consider sustainability, or as Christopher Spencer describes it 'green is good for you'. While the more active dimensions of sustainability – such as the incorporation of photovoltaic panels to generate electricity – have value, are they affordable and efficient enough to warrant the initial financial outlay? Of more value are the choice of materials. Are they low in embodied energy? Can they be recycled? Are they natural materials? Is there a green view? What does sustainability actually mean to you?

3. Children move around the environment much more than adults do. The pattern of young children's play, as opposed to older children's, is often characterized by free flow from activity to activity, transforming as it goes. The clear implications for the design and running of the kindergarten is to foster and to afford this fluidity by providing points of play throughout the nursery's spaces, not just within each activity area.

4. Can the building evolve and change in line with its users? This is a similar idea to the above, but it is an idea about the transformation of the environment in ways that the architect may not have initially recognized. The ability for the environment to be adapted during the following years of its use is the mark of a true learning environment.

5. Are the hard qualities, such as safety and security, balanced by the soft qualities, such as low seats, curtains and colours which help to make it truly child oriented, considered in equal balance? If there is too much hard stuff, the building will not help to stimulate the child to learn.

6. Complexity is good, albeit at an optimal level, as it helps to encourage exploration and stimulates interest (too much will leave the children bewildered and lost). Provide opportunities for exploration by adding corners and hidden areas. Sometimes the design can just happen when it relates to its site and what is already there. Vary ceiling heights in slope and fenestration, for example.

7. Consider evidence from other disciplines, such as environmental psychology. Dare to dream and spread your own personal frame of references. Too many preconceptions and an emphasis on health and safety concerns will make the nursery a dull place.

Chapter 2

The sustainable nursery

A good environment is a natural environment

The degree of change experienced by the past three generations rivals that of a species in mutation. Today's 'screenager' – the child born into a culture mediated by the television and computer – is interacting with his world in at least as dramatically altered a fashion from his grandfather as the first sighted creature did from his blind ancestors.[1]

Making the case for nurseries

As a designer largely specializing in early-years facilities, I am often asked by client organizations why so much money should be spent on new buildings to provide a particular child-oriented environment. Sometimes they will cite their own experiences as small children, perhaps having attended the local playgroup 20–30 years previously, where very few qualitative modifications to the environment were made on behalf of the children. Their memories probably hark back to that time when children and their families made the best of an early-years care system that was informal and home-based, run largely by non-professional carers, predominantly directed towards a socio-economic culture of the full-time stay-at-home mother. Those who were poor or socially excluded had nothing at all.

Over the past few decades, childhood has changed. It has certainly become dominated by a pervasive digital culture which barely existed 25 years ago. Even three-year-old Amy is starting to use computers, with sophisticated programs aimed directly at her age range. She and other children in her nursery will grow up in an environment where globalism rather than localism predominates. This has profound effects, not least because global warming as a result of the consumption of fossil fuels at an accelerating rate, may make 'The Ark Floating Nursery' a practical model for the future of our children.

There is a further dimension to this story. That is the relative lack of freedom afforded to children in the largely urban environments of our modern towns and cities. This might be compared to previous generations where children would quite literally disappear from the home and exist on the streets and in the fields around home, playing largely unsupervised with other children (during the time when they were not at school). Colin Ward, in his classic book first published in 1978, *The Child in the City*, describes how children used city spaces as venues, hidey-holes for secret games and for sports as various as fishing and ferreting.[2] Back then, childhood was played out in public places generally away from the immediate control of parents and primary carers. It was more social, with intensive interaction engaging other children and also with adults usually on an informal basis.

Today, being on the streets usually labels young children as uncared for or coming from poor, uncaring families. Most adults would hardly dare make eye contact with children which are not their own for fear of being labelled as an interfering busy-body or worse. Parental control and supervision to ensure total safety from a media-induced plethora of dangers, real or imagined, makes the modern child somewhat confined, lacking the freedoms enjoyed by previous generations. Children, it seems, have almost disappeared from public view. The pleasure in watching their unselfconscious, exuberant games – even the annoyance of their taunting and chasing each other and dramatic emotional extremes – is now all but hidden from view. Children are perceived to be in danger outside their own domestic space, and to be dangerous to others if they roam unattended:

> The answer about why today's children cannot have fun without parental supervision is simple. We had one quality missing in the life of children today, freedom. Provided we returned to the nest at the agreed time, we could go where we wished and thereby develop our creative and imaginative skills without the need of adult help/or sophisticated toys. Now we are obsessed with protecting our children against traffic, abductions, molestations, mishaps on school trips, drugs – the list is endless.[3]

So, 20–30 years ago, while children's lives were much less newsworthy (unless something really dreadful happened, as sometimes it did), children were much freer to play and explore on their own. Therefore the need for nurseries and other purpose-designed children's environments seemed to most people to be of a very low priority. Unstructured spontaneous play was simply everywhere, an activity that was acted out in public as opposed to the private places like the home, where it largely takes place today. These places were sometimes visible to passing adults – for example, street games – or they were partly or wholly hidden from view, such as the boarded-up bomb sites to be found in Britain's major industrial cities in the 1950s and 1960s. Behind the hoardings children could play relatively isolated from adult view. By and large, children found their own places to play, which were generally outdoors and unconfined, unused spaces in towns and cities or fields and hedgerows around the suburban sprawl. Children's play areas in parks and other pre-planned public play spaces were developed by Local Authorities widely during the post-war years. But for most children, they were generally to be avoided, lacking as they did the magical qualities of the secret unplanned play places, bomb sites, abandoned cars, underground dens, streams, left-over spaces which have today all but disappeared.

While the design of conventional play equipment is better than it was 50 years ago, it has limited sustained appeal to most children compared to natural phenomena such as this tangled tree trunk in Richmond Park or the beach at Cape Cod, where children make real sensory contact with powerful natural phenomena which are both exciting and risky, thus double the fun.

Consequently, most spaces dedicated to childcare or family support in those days would have been 'as found' rather than adapted or purpose-designed to the needs of the users. Usually it would comprise a single multi-function open-plan area such as a church hall or a community centre which would be used by the local scout group in the evenings. Aside from critical child protection issues such as safety, security and hygiene (children can't play freely when the floor is filthy), these environments would hold little interest to children – hardly stimulating and in fact profoundly boring them sometimes to the point of disorientation. Staff and carers would trundle out toys which would be returned to their boxes at the end of each session. Play was confined to adult dimensions. The building hosting these activities had little purpose-made value for the children in question.

The sustainable nursery

An adult and child working together in the outdoor art room at Pacific Oaks.

Inevitably, because the environment limited the potential for extended imaginative play, the effect would be to depress the child, make staff and carers work particularly hard at the task of keeping children happy and stimulated, thereby restricting, almost excluding, the involvement of parents in the lives of their growing children. Such facilities would be virtually impossible to provide for their needs when catering for full, morning to evening day-care. It was not surprising, therefore, that early-years daycare was largely viewed as a second-rate place for growing children, suitable only for vulnerable or at-risk children; the orphanage was a term used to frighten children, such was the barren, austere nature of the regimes and the environments which supported them. Planned state provision of early-years care in the United Kingdom was, until relatively recently, a concept which had very few values in terms of its child-centred qualities.

This was certainly not the case in Scandinavia at this time, where full daycare was very much more integrated into the education system, infused with an instinctive and demonstrable appreciation of children's lives and children's culture generally, which usually made the nursery a place children actually wanted to be. In Sweden there was a massive explosion in nursery school provision during the 1970s, when the manufacturing industries were expanding, causing labour shortages which were to be filled by the large-scale entry of women into the industrial workforce.

In Denmark the system was also very coherent, having developed from a fundamental sympathy and understanding of liberal, cooperative socialism. Crèches provided full daycare for babies up to three years old. Public kindergartens provided care with informal educational activities for children from three to six years old, when compulsory school starts. Right now, age-integrated institutions are also being developed which attempt to cater for the needs of children from birth up to 14, an ideologically driven process to erode the synthetic division of children into socially segregated 'squads' based only on their age. On a social level one asks the question, if we are trying to give children the capacity to relate to an adult world when they themselves grow up, where within the adult world do we find age-determined social situations? The key pedagogic implication being that children learn as much of value from other older children as they might from sparse access to adult carers during their day in nursery.

There are, however, many age-integrated nursery schools for children up to six years old in the Scandinavian system, whereas in the United Kingdom separation is still the predominant organizing principle. In addition, there are so-called *fritidshjem* or centres for school-age children, which are often part of the kindergarten itself. These cater for the needs of school children in an elementary school system which terminates at 1 p.m. or 2 p.m., a common model throughout much of Europe. Although there are some private kindergartens in Denmark, 65 per cent of children attend state-funded crèches and 76 per cent of the three-year-old population attend publicly funded institutions of some sort, many on a full-time basis. By comparison, in Britain only 41 per cent of three-year-olds attend.

There are important differences between the essential nature of the Scandinavian pre-school systems and the Anglo-American model, which is rooted in a more entrenched political philosophy. When Danish educationalists refer to the term 'social pedagogy' (loosely translated), they are not alluding to a political idea as such; in this context the term 'social' refers to the primary goal of the kindergarten or nursery, which is to introduce the child to social interaction, particularly with other children, and to recognize that the child has rights which must be respected. This is almost analogous to the right to enjoy their own public realm, which means that common values and willingness that people get along is the overriding requirement of early years. 'Education' as distinct from socialization is perceived in Denmark as something aimed at older children, above compulsory school age – six years old or more. It is clear, therefore, that the practical way in which Scandinavian thinking can be realized is to allow children to construct their own activity patterns and games rather than having adults dictate to them towards some spurious pre-conceived educational outcome. Adults may be catalysts to activity, but beyond that they are merely role models and enablers. Therefore the Danes do not refer to 'nursery school teachers', but to 'pedagogics'. The concept of a separate and distinct profession of early-years leaders, enabling tasks and activities directed to the pedagogic goal, is sometimes hard for us to understand, particularly as we lack a longstanding tradition of widely accessible early-years services.

In the United Kingdom over the past ten years there has been something of a social revolution in the entire organized childcare sector, which has affected perceptions of early-years care and education for the better. The in-coming Labour government of 1997 saw and recognized the isolated condition of many impoverished families and their children in the United Kingdom, particularly in the de-industrialized English and Scottish heartlands. Thatcher's failure to recognize the need for early-years care (to paraphrase her view, my children did not need it, so why should others?) can be mischievously set against the result of her policies during the 1970s and 1980s in this country as millions of people lost their jobs, an effect which has created entrenched welfarism and longstanding deprivation in many communities, where sometimes three generations of a family have never worked. As a catalyst for the introduction of child and family support based partly on the European model of municipal nurseries and partly on the American 'Headstart' system, it was a no brainer.

Social conditions relating to women's rights had also changed profoundly. The view that it was damaging for young children to spend time away from their mothers or primary carers in the early years was a commonly held belief articulated in the United Kingdom by the psychiatrist John Bowlby (among others) during the 1950s. This created the male-breadwinner–stay-at-home-mother model of that time. This might be fine if the family could afford it; its effectiveness in poorer communities depended on strong local family or friendship ties which prevented social isolation, which is so prevalent in these communities today.

Extensive research carried out during the 1970s and 1980s largely disproved Bowlby's theories of attachment (that children would be damaged by separation from their carers), provided that the alternative nursery care is of good quality. However, the women's movements of the 1960s combined with the introduction of the contraceptive pill in 1964, which gave control to women as to when and with whom they would become pregnant, probably had a more profound effect on the drive towards more nursery provision. Certainly, from that time many young women were no longer prepared to commit themselves to parenthood during their twenties, preferring instead to carve out a professional career instead, deferring motherhood until their thirties, and today for some even later. As a byproduct, the fertility rate dropped from 2.94 children per woman in 1964 to a record low of 1.63 in 2001.

Many professional women demanded to continue their careers after childbirth. At middle-class dinner parties up and down the country, discussions focus on childcare, much as it used to do in Scandinavian countries during the 1960s and 1970s. The options are usually stark for working mothers, as private nurseries can consume a fair proportion of a wage by anyone's standards, just to provide a level of care which is safe, nurturing and vaguely educative. Today, economic necessity combined with certain social pressures to earn more and to benefit from a higher standard of living means that two parents are usually working full-time. With weakened family ties and more single-parent families, childcare I suspect is more of a necessity to support new patterns of working than it is an educational catalyst for the development of children; although many parents recognize the social benefits of early-years education, in the overall scheme of things this is probably less of a consideration than the need for someone to look after baby while mummy goes out to work, from early in the morning until the evening.

For parents in deprived communities up and down the country, those who are unable to work or choose not to, government policies now recognize the particular hardships faced by the children living within these straightened circumstances. Current policy aims to support these children with a more comprehensive Nordic-style early-years system which combines the best mixture of care and learning led by properly qualified teachers who have an affinity with the distinctive needs of early-years children. If they do not have some form of early intervention and support, it is feared they will miss out on life-enhancing development structures and positive role models crucial in their early years. It is clear that children from poor families will be at a significant social disadvantage to those who attend some form of good-quality early-years facility. These disadvantages may last a lifetime.

This will also have wider effects on the quality of public life, breeding more dysfunctional adults at odds with society as a whole. This is a problem that is more marked within societies where extreme levels of wealth sit adjacent to extreme levels of social deprivation, (long-term unemployment) and poverty. However, as I write (and since I made the observations above), the so-called global credit crunch has kicked-in and the United Kingdom and United States are experiencing a profound and painful period of economic recession and all that goes with it. Within this context it is fair to say that the engagement of adults within the structure of childcare support is as important as it has ever been. The model of the children and family centre which includes rooms for adult training as well as nurseries for children is being developed at some pace in England at present. It follows a tried-and-tested model from European countries such as Italy and Spain.

It may appear to be a somewhat didactic depiction of nursery care within the United Kingdom, the middle-class need to find life's fulfilment through work, set against the working-class need to shelter uncared for children. However, this is the polarized condition in which the debate is usually conducted. The myriad shades of grey that fill-in between the extremes described above makes simplistic definitions of what is required hard to unpick. It is clear, however, that childhood is now a highly politicized area of widespread interest and concern which is important to all stratas of society. However, beware politicians looking to fix childcare on the cheap – it has to be a long-term commitment which has a strong hint of socialism about it, or to put it another way, nursery is integral to the 'big society'.

The sustainable nursery

This is not a systematically researched publication and it is beyond the scope of this study to justify the need for daycare on the basis of straightforward scientific analysis. Besides, while building science can provide accurate records of physical parameters such as illumination or temperature, relating these factors to people's expectations is difficult. Therefore this publication is to a greater or lesser extent a subjective view. However, it is structured around 25 years of experience working within the field of early-years environments, with a lot of observation, thinking and reading around the subject in the meantime. And of course the experience of nurturing my own children as a single parent should not be underestimated in framing my positive view of nurseries.

It is fair to say that I am passionate about the benefits they hold for the world that my own children will inherit, and on a purely selfish level a good nursery has given me the time and space to be able to sustain a career (of sorts). However, some justification is required because there remains a stubborn and often erroneous debate around the pros and cons of daycare, and in particular the negative effects some young children experience when separated from their parents at an early age. I often get my evidence of need from the popular press, as one new expert or another comes along with their heartfelt views of childhood, usually around the time of an unfortunate childhood fatality, abduction or cases of criminal parental neglect.[4] But mostly I get my ideas from firsthand experience, visiting children's centres and discussing issues with staff and parents, designing and building children's centres myself as a practising architect.

Some evidence suggests that children who spend extensive amounts of time away from parents, more than 20 hours per week in the first year, are likely to show insecure patterns of attachment. Certainly it is, in my view, a truth of such logical sense, that it hardly needs stating. The care of babies in particular is an area to which I will return later. However, when children are a little older, aged 18 months onwards, patterns of attachment are usually less dependent on their parents and they can be equally supported by expert primary carers.

However, other evidence from the field of psychological studies indicates that daycare children usually out-strip their non-daycare peers on measurable levels of sociability, persistence and achievement.[5] Not at all surprising you might think, bearing in mind that in a well-run establishment, a daycare child interacts continually with children of his or her own age and this social interaction is fundamentally enhanced by the building itself, particularly where the natural movement of young people around the building is promoted by its planning and the organization of its spaces. In this respect, the design of the entire nursery is absolutely fundamental. An example of this is shown in the plan of Cherry Lane Children's Centre: the building is conceived as a series of child-oriented 'events' which roughly correspond to the movement patterns of young children. Observational studies show that they constantly move around from event to event, if the building is planned according to these principles.

Not only will this enriched social experience go some way towards compensating for what is lost in the diminished contact the child has with their own family, evidence also supports the view that good-quality daycare

The sustainable nursery

List of the wide selection of play sessions and events available through this North London Children's Centre. Social activities are important.

SESSIONS AND ACTIVITIES RUNNING FROM JANUARY 2007

MONDAYS	TUESDAYS	WEDNESDAYS	THURSDAYS	FRIDAYS
10.00am – 11.00am Sure Tots Walking + @ St Augustine's, Bull Farm	9.00am – 11.00am Child Minding Group @ Pleasley Hill Children's Centre	Big Cooks, Little Cooks @ Pleasley Hill Children's Centre, coming soon	9.00am – 11.00am Child Minding Group @ Pleasley Hill Children's Centre	9.15am – 11.15am Stay at Play @ Sutton Road Primary School
10.00am – 12 noon Let's Talk @ Pleasley Hill Children's Centre 22nd & 29th Jan & 12th Feb	9.30am – 11.30am Drop in Creche with family support workers @ Ladybrook Children's Centre until 23rd Jan	9.15am – 11.15am PEEP for Ones @ Ladybrook Children's Centre until 24th Jan	9.30am – 11.30am Communication Skills @ Ladybrook Children's Centre from 4th Jan for 4 weeks	9.30am – 11.00am PEEP for Babies @ Pleasley Hill Children's Centre
1.00pm – 3.00pm Stay & Play @ William Kaye, Ladybrook	9.30am – 11.30am Building Blocks Understanding Children's Behaviour @ Ladybrook Children's Centre from 30th Jan for 5 weeks		9.15am – 11.15am PEEP for Ones @ Ladybrook Children's Centre from 1st Feb	11.30am – 1.00pm Baby Café @ Pleasley Hill Children's Centre
1.00pm – 4.00pm C Card drop in Free contraception for under 19's @ Pleasley Hill Children's Centre	10.00am – 12noon Stay & Play @ Crescent School	1.00pm – 3.00pm Family Support Drop In @ Pleasley Hill Children's Centre	9.30am – 12noon C Card drop in Free contraception for under 19's @ Ladybrook Children's Centre	1.00pm – 3.00pm Young and Proud Parents YAPP @ Ladybrook Children's Centre
1.30pm – 3.00pm PEEP for Ones @ Pleasley Hill Children's Centre	1.30pm – 2.30pm Sure Tots Walking + @ Pleasley Hill Children's Centre		1.00pm – 3.00pm Stay & Play @ Pleasley Hill Children's Centre	1.30pm – 2.00pm Rumble in the Jungle 0-18 months @ Ladybrook Library and also at 2.15pm – 2.45pm for 18 months - 4 years
			1.00pm – 3pm Peep for babies @ Ladybrook Children's Centre	

promotes independence and self-sufficiency from an early age. My conclusion is that, overall, when daycare is good, children do not suffer; however, children from disadvantaged home environments are most likely to benefit rather than be hurt by daycare. Here, evidence suggests that those benefits are significant in terms of raising low expectations about their own life chances when they come from an unnurturing home. Already, findings from Britain's Effectiveness in the Provision of Pre-school Education research shows how much better children do at the age of ten in reading and maths if they have had high- or even medium-quality nursery education. But this evidence will not show for a long time, even a couple of generations down the line, when today's nursery-goers are in a position to reflect on their childhood experiences.

Some evidence suggests that when the quality of care is poor, infants enrolled in daycare centres end up inattentive and unsociable in school, compared to children who spent the same amount of time in good daycare centres. In a similar study, four-year-olds who attended higher-quality daycare centres showed better social and emotional development at age eight than did children who attended poorer-quality centres, even when factors such as social class and income were equated. However, broadly agreed definitions, which transcend professional disciplines of what constitutes 'poor quality' and 'good quality', are hard to come by. Therefore one of my aims here is to establish clear benchmarks for what constitutes good quality. At the end of each chapter the reader will find a systematic designer's checklist of features and considerations. It represents a synthesis of the author's previous observations and research, taking into account as many as possible of the comments received from interested parties with whom I have debated long and hard on this. It is not cast in stone; rather, it should be viewed as an evolving theory which aims to prioritize issues in a systematic form.

The sustainable nursery

The influential study carried out in the United States over a period of 30 years into the HighScope/Perry scheme (under President Lyndon Johnson) is often cited as evidence to support the argument for comprehensive, good-quality childcare. During the research a huge amount of support was offered to one sample group of disadvantaged families; as well as daycare for pre-school children there was extra tuition for school-age children, therapeutic and practical support for mothers and fathers. In short, a wide range of welfare benefit (but not money) was provided which was directed towards children and families. Nothing special was done for a comparison sample of families, and 30 years later it was found that the HighScope graduates had done dramatically better. For every dollar spent on HighScope, the economy was seven dollars better off further down the line it concluded. Money was saved on police, prison and other social services, and gained from taxes on earned income and unclaimed benefits. Not only did the HighScope group avoid crime, they were educated, got jobs and contributed to the economy.[6]

What makes a good nursery and how do we raise quality? Smaller group sizes and age-differentiated groupings common to some Montessori institutions are just two ideas which will help to alleviate these problems. More space than current guidelines prescribe is almost certainly a critical factor in all of this, including the need for high-quality outside space as is the norm in Scandinavia. Of course, all of this has significant economic implications as well. A high-quality environment is important. However, it is fatuous to suggest that architecture alone can straighten-out dysfunctional or socially alienated individuals. St. Pancras Station may be a beautiful building, but it won't make the trains run on time!

However, environmental quality and what it provides in terms of play opportunities is one important dimension of this quality agenda. The more diverse and richer these play opportunities are, the better that environment will be. An extreme example of a nursery with poor play opportunities is provided by my own son: Matthew attended a large corporate workplace nursery from the age of 2.3 years – admittedly this was some 15 years ago. His abiding memory of that place was the sheer boredom of it all. One of the key toys for boys to grab was the pink hairdryer, which could be used as a gun – there wasn't much else. This odd deprivation at the hands of such a wealthy corporate body as the BBC has, I am sure, helped to inform his longer-term view of education as negative.

Another key dimension in defining quality is surely the social patterns which the best institutions implant within the young mind. An example is the eating ritual that I have witnessed in so many Italian settings, where lunchtime is a group activity or event around which activities of the entire day hinge, embroidered with good-quality fresh food and drink, with no options to confuse the young child with the notion of choice; food is food and the activity is pure and simple, pasta or bread, meat or fish, water, fruit, yoghurt, all fresh and organically produced. However, the design of the restaurant is at the same time environmentally and spatially complex, with important features such as the daily menu posted at the entrance so that parents can see what their children are eating, the solid adult-sized beech table where both adults and children of mixed ages sit together, some in high chairs replicating the home arrangement with set places established from the outset, each of which belongs to a specific child. The invitation to 'sit in your place' is unambiguous and deliberately individualistic, thus establishing from the outset the immutable consistency of this activity within the clear social framework of a group setting. Its practical purpose is one element of the activity, the social and symbolic role it plays is of equal importance and is clearly a spatial requirement of the nursery.

This can only happen because it is what the environment allows; comparing this to so much eating in UK daycare settings, where temporary places will be set-out in the play/activity areas, with poor-quality 'fast' food served up because it is cheap and easy to prepare (cheap meaning it is mass-produced and stuffed full of harmful 'E' numbers so that it can be re-used), which somehow undermines this activity as a social norm to be valued and enjoyed throughout the life of the child. It tells the child that he or she doesn't matter, rather than implanting more positive messages.

Inevitably, this is a complex cultural debate, and in the less regimented more haphazard market-driven societies such as England and many American states, it is difficult to suggest that these types of regimes could ever be imposed upon diverse peoples. However, there is no doubt that earnings inequality and child poverty within society are the single most important factors which continually undermine the well-being of children in early years at home. Early-years settings work well when happy, rested, healthy children attend who are ready and capable of behaving well within a cooperative environment. Unfortunately, the unequal societies we have created in the United States and England are far away from the norm in many European countries such as Italy and Scandinavia, which have proven to be longstanding models of good early-years practice, which too few of our politicians currently understand or even recognize. However, as previously discussed, the global credit crunch may well bring about changes to the individualistic basis of society since the Reagan–Thatcher consensus of the 1980s.

It would be interesting to check if this negative research applies to children brought up in the high-quality Italian models, such as Reggio Emilia. In addition, it is an overwhelmingly persuasive argument to state the benefit a child gains from no longer living in poverty if his or her single mother is in work. One of the most touching stories I have heard was the three-year-old child with drug addicted parents who got herself ready in the morning while her parents slept. She turned up at nursery alone to have breakfast and enjoy the rest of the day in Sure Start. Whether or not that child will end up being more aggressive is a moot point; what is clear is the need to support the immediate needs of the most deprived and needy children in our society such as this lonely individual (who should in a supposedly equal society have the same life chances as any citizen), despite the possibility of negative effects on social interaction further down the line. It is a struggle to persuade many people of its value. However, as Stephen Kline states, 'A civilized society is one which struggles to make the world better for its children.'[7]

The difference between much run-of-the-mill 'nursery' provision and nurseries with a sophisticated and well-delivered curriculum such as High-Scope may explain a great deal about the variable outcomes of this value-laden research. A successful and effective nursery is so often to do with the

In the Scuola dell'Infanzia Diana, Reggio Emilia, playing with shadows on a light box. The Reggio Emilia early-years system is generally considered to be the best in the world. Well funded by the regional government, the concept that children come first is a city-wide principle which can be seen not just in the nursery buildings and the pedagogy which goes with it, but also in the wider urban planning with dedicated play streets and safe cycling zones throughout the city of Modena. Ironic, you might think, since this is the home of Ferrari.

Here the environment is described as the 'third teacher', which relates to details which promote play. For example, they have discovered the importance of using unusual light sources such as overhead projectors and mirrors: 'One characteristic that should be accentuated is a type of mixed or virtual indirect lighting, in which the source is rarely visible and the light all seems to be indirect. In reality the light is both direct and indirect, but the two types cannot be clearly distinguished because of the use of filters and the numerous reflections and micro reflections of the "second skin" of materials and communication panels that cover the walls, ceiling, table and objects.' (Reggio Children, *Children, Spaces Relations*, Domus Academy Research Center, 1998, p. 55).

nature of the curriculum and skilled, caring staff able to deliver is a reality which may go a long way towards explaining the profound differences between the two research outcomes. HighScope is a very particular approach which relates specific activities to the environments within which they take place. This contrasts with the less structured, more careless systems of much early-years activity. What is particularly distinctive about HighScope is the essential engagement with the environment in every aspect of children's play. For me this marks out HighScope and other intensive early-years curriculum programmes as the key ingredient of subsequent success in daycare. Inevitably, the quality of the environment is a fundamental component in all of this. However, the capacity of children to engage is also profoundly important. In this regard, the child's self-discipline is primarily a factor of staff and parental support. Without a degree of control, no matter how good the environment, very little value will be gained.

HighScope and some children's centres discussed

The HighScope approach evolved during the 1960s as a result of work undertaken in the United States with children from deprived neighbourhoods in Ypsilanti, Michigan who were at risk of school failure. It was felt that the children lacked the opportunity to develop intellectually, rather than being inherently of lower intelligence than other children as some might believe.

Because of a lack of storage space, nursery interiors often seem to be full of clutter, bursting at the seams to the point where it is difficult for staff and children to move around freely. The shop at London Zoo provides children with a much more organized and legible internal landscape (for cynical commercial reasons, it must be stated). With toys displayed in this ordered, accessible way, the architecture becomes attractive and complementary to the 'second skin'. Designers should ensure that 25 per cent of the overall space is given over to storage, with as much of it open to children as is practical. Clear pathways for children should be considered by both architects and play managers.

The sustainable nursery

The aim was to provide a structure for their early-years activities which enabled rich interactions with adult carers and, as previously mentioned, with the environment itself. Based on the child development theories of the Swiss psychologist Jean Piaget, HighScope aims to engage children in direct, hands-on experiences with people, objects, ideas and events.[8]

At its core, the system assumes that children gain confidence, initiative and the love of lifelong learning when they are involved in well-supported activities of their own choosing. It views children as active learners who prosper by being engaged in activities that they themselves plan, with the aid of a HighScope-trained teacher, and a suitable environment. The more supportive is the environment, the more progress will ensue. They can then carry out and reflect upon their activities, interacting with their environment at every point along the way. Indeed, because the idea of self-initiated activities is the key to this approach, HighScope can only happen in a place where the 'tools of play' are readily available. Therefore the way in which toys and other play activity materials such as paint, paper, glue and scissors are stored is an essential part of the architect's brief. The childcare environment becomes the forum for child activity. It is an integral component of successful development.

It is worth describing the daily routine which structures a typical two-hour HighScope session in more detail because it relates so intrinsically to the environment available. Note how each period of activity is precisely defined by the period of time in which it is supposed to take place. First, there is 'greeting time', which is intended to provide a smooth transition from home to school. This is followed by the first event, 'large group time', which lasts for approximately 15 minutes. This comprises children and adults playing games together, telling and re-enacting stories, singing songs, doing action rhymes, dancing, possibly even playing musical instruments (on a basic level) or the re-enactment of social events. It gives adults and children a chance to share important information for the day, emphasizing group collaboration at all times. This is followed by another 15-minute period, which is described as 'small group time'. This period is more adult-initiated, where materials are presented and discussed. Children themselves then decide how they will use them. The next stage is 'planning time' (ten minutes), when children indicate what they plan to do with their work time, the materials they may use and possibly who they will work with or alongside. Adults will talk to and listen to each child individually. 'Work time' then follows, which is a longer session, possibly of one hours' duration, where children carry out their plans, working with any materials, indoors or outdoors. Adults look for opportunities to enter into the children's activities, to encourage and extend their thinking, to promote their play patterns towards more sophisticated models of activity where interaction with other children is constant and to enable them in problem-solving situations, but never to do it for them. There is inevitably a great deal of adult supervision in the early stages of a child's attendance in the form of imposed structures of discipline and care, which ensures that children 'stay with it'. In practice, when the child is used to the routine, they become remarkably self-sufficient. Again, we stress the importance of the building to provide for both fully supported group time and self-sufficient individual or small-group play.

'Tidy up time', which takes 15 minutes, comprises children and adults working together to return materials to their proper place, including sweeping the floor, cleaning tables and washing their hands. Again, this emphasizes the importance of controlled discipline within the environment. 'Review time' then takes place and is perhaps the most important activity period, where children are encouraged to reflect on and talk about the work they have done. This lasts for at least ten minutes. 'Snack time' then follows, with milk and fruit shared over a 15-minute period, usually around group tables; this is followed by a final 'large group time'. This is another ten-minute period of discussion and interaction with more songs and stories. Finally, parents either collect their children, or for full daycare children, they will go to another quiet part of the nursery to sleep or rest over lunch.

HighScope is often criticized for being overly structured, placing a strain on children and carers to be constantly 'at work'. Many educationalists

The nursery has all resources on display and generally accessible to the children; it is 'ready for play' (below).

Organized activities such as the morning snack-time (above right) are important for the development of social skills. Tidy-up time (right) is valuable for children for obvious reasons.

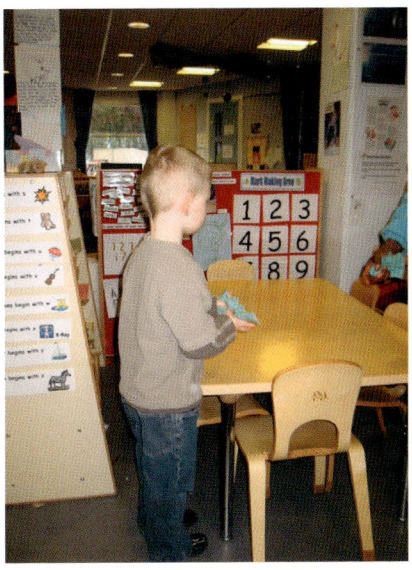

look to the possibility for children enjoying more independent activities on their own or with one or two other children, and in a less frenetic way. However, its structured discipline is a strategy which somehow flies in the face of the 'anything goes' freedom of current times. It should be stressed how important it is to give children the opportunity to reflect and do nothing if they so wish, particularly over the long days in a daycare setting. I once visited a centre in Denmark and was concerned about a child who was just sitting under a table staring into space. Asking the pedagogue about this he explained that they saw it as a positive period of calm and reflection for this particular child, who merely wished to do nothing, but within the context of the normal activities going on around him. Often it seems that we need to fill our young children's time with activities, never leaving a single moment for individuals to be quiet and reflective. This would be an obvious critique of HighScope.

It should be borne in mind that while this is similar to a primary-school system in its strict narrative structure, it is only intended to last for two hours, rather than being a full daycare form. It is also clear that much childcare is based on varying degrees of structured activity; only childcare of the poorest quality ever leaves children on their own to do as they wish.

Certainly there are justifiable concerns about overloading young people with structures and systems which are too predictable. However, the counter argument is that even adults will feel unsafe and unsure when 'anything goes'. The HighScope curriculum offers a structure to the day which provides a predictable routine within which the child and the teacher can feel secure and develop their own sense of self-discipline.

This has an important effect on design. For example, materials and toys must not be hidden away in cupboards with doors, or in closed off storerooms. There must be an order to the classroom spaces, but the means by which the children explore their interests must be open and accessible to them at all times. It is a compelling idea, which does not always sit well with the prevailing architectural concept of tidiness and aesthetic control over the environment. As mentioned previously, early years is a very particular architectural challenge. Understanding that there is a curriculum and that space is particular to the special activities which take place there is important knowledge and profound understanding for the designer. Prosaically, it may be that one of the architect's key questions to his client users is going to be 'how do you store things?' so that the child has his or her own access to them at all times.

Over the past ten years within the United Kingdom, many of the HighScope principles have been implemented consciously or on a more intuitive child-centred basis, both within the private sector but more visibly through the children's centre programmes. In 1997, the new incoming UK government started to develop their plans for a coherent system of child support for pre-school children. At the heart of this policy has been a commitment to construct high-quality, purpose-designed children's centres as an essential element of this support system. The key principle was the recognition that childcare was no longer a service that could happen organically in any old left-over bit of space. It needed purpose-built environments and a clear structure of revenue funding and governance to create appropriate places which would be financially sustainable. Places within which the

There is a stark contrast between the 'LITTLE CHERUBS KINDERGARTEN' façade and that of the 'kindergarten Jerusalemur Strasse', Berlin. One is a barely disguised garage conversion, the other is a multi-million euro celebration of children. Each expresses the value we as a society place on our children's care and well-being.

services could be dispensed not just for children but also for adults were considered as being important, particularly in areas of high unemployment. There is in my view also a profound symbolic message here, which is often overlooked. It recognizes the social importance of early years, so that new parents can meet other new parents and share the pleasures and the pain of this life-transforming event. Any institution is identified by its buildings; after all, you can in theory have a religion without a church, but in practice it is virtually unheard of. Defining quality thresholds, including the possibility of symbolic messages being communicated within this new building type, is the main task of this publication.

One of the key issues in the UK system is I believe the lack of an early-years culture and tradition. This is not to say that early-years teachers and care workers are inferior to their European counterparts – far from it. It is more to do with those who use the services, a generation of parents who did not themselves benefit from early-years care and education, who as a consequence generally lack the knowledge and understanding of its systematic approaches. Its principle of discipline within the framework of love and tenderness, the importance it places on social interaction and the value of spatial awareness are concepts which parents and teachers by and large do not 'get'. If all daycare was of a consistently high quality, the debate could, in my view, be laid to rest. However, much is not; factors such as adult–child ratios, how well teachers are trained, selected and paid (dictating staff turnover), the understanding and empathy of parents who use it, and most importantly for us here, how these key users enable the children to relate to the physical facilities, are the key components in the quality agenda.

A large-scale US study carried out in 1995 concluded that 86 per cent of daycare centres qualified as poor or mediocre environments, with only 14 per cent qualifying as good. Similar standards applied in the United Kingdom around this time. The picture has improved immeasurably over the

The sustainable nursery

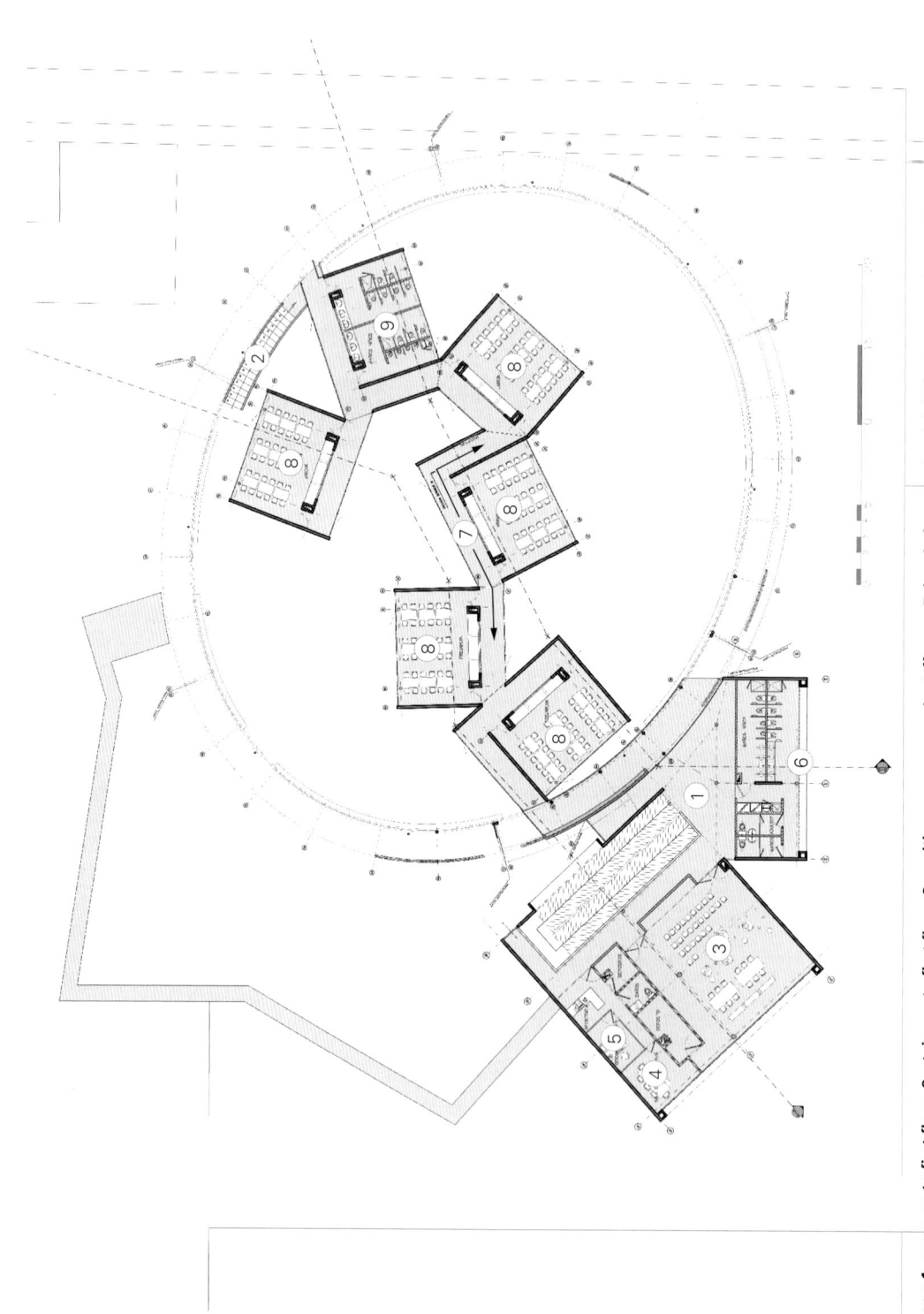

First floor: 1. ramp up to first floor; 2. stairs up to first floor; 3. multi-purpose room; 4. staffroom; 5. head of centre; 6. adult WCs; 7. connecting walkway; 8. classrooms; 9. main child WC and wet area.

The sustainable nursery

Ground floor/first floor: 1. entrance; 2. kitchen; 3. laundry; 4. canteen; 5. adult WCs; 6. nursery; 7. nappy changing; 8. classrooms; 9 patio courtyard; 10 entrance court; 11. ramp upto first floor; 12. child WC and refuse store; 13. child WC and wet area; 14. buggy store and reception.

49

The sustainable nursery

Often, financial expediency acts as a catalyst to new and innovative approaches to design and procurement. This building, the El Porvenir Kindergarten by Giancarlo Mazzanti Architects, is for an impoverished community on the outskirts of the Columbian capital, Bogota. It uses pre-fabricated modular units for the activity areas, which are sprinkled across the site seemingly at random like giant building blocks. They are interconnected by a kinked circulation spine to create a single coherent unit. An oval colonnade/wall structure wraps around the entire compound, forming a secure yet architecturally complete composition which has become a centre for the community since it first opened in 2009. It is practical, being secure and economical to build, yet it adopts an open architectural language which enhances the building's wider social mission; for example the lattice wall is protective yet permeable.

The oval outer wall is a two-storey circulation route which also forms a defensive enclosure for the kindergarten external play-spaces and classroom 'houses' within.

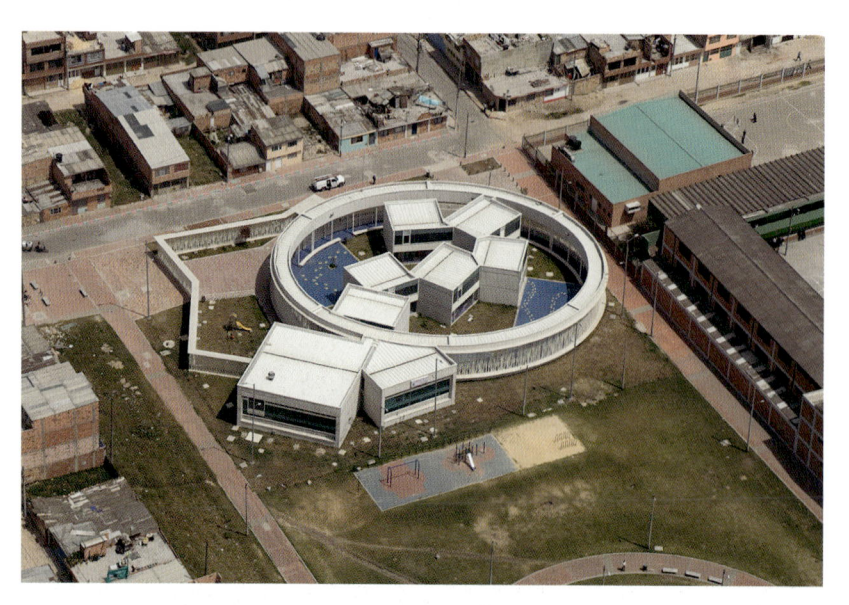

This aerial view shows an organizational clarity which carries with it a level of complexity which is both intriguing and architecturally legible at the same time; inside, the children's houses are distributed in a higgledy piggledy form, creating a jumble of different external play areas. The adult-oriented accommodation outside the oval is more ordered, relating to the solar geometry and the surrounding urban grain.

The sustainable nursery

Children's activities within the oval play area, gently shaded from the morning sun by the giant two-storey enclosure. When it is very hot, activity sessions take place in and under the colonnaded walkway, with the long, curved stone bench forming an appropriate child-level play surface for reading and block-play.

Another view of the children's play yard behind the WC/wet area.

The sustainable nursery

With its contrasting primary forms and distinctive white filigree colouring, the new building contrasts with the surrounding urban grain to present itself architecturally as a vibrant new centre for family life in the expanding outer residential suburbs.

intervening ten years; however, far too much is still dictated by the pressure to build cheaply and quickly, often at the expense of the long-term well-being of the most vulnerable members of our society. Of equal importance is the value of staff: pay them poorly and restrict their own potential for growth and self-improvement, and quality will also be affected. They will tend to move on. Stable institutions are good for children, unstable ones are bad.

In the commercial world, resources are lavished on the environmental comfort of adults; we should go some way towards providing an equally appropriate level of environmental quality for young children and their carers. Buildings should not merely satisfy basic needs, they should provide the right amount and type of space for activities, which will be of interest to and stimulate their users. Most good architecture combines the practical with something less tangible; a sense of delight in the spaces which make up the building as a whole. At its best, refined environmental qualities may even hold the power to modify the moods of its users in a positive way.

For example, research in the United States into elementary school classrooms shows that students in classrooms with the most daylight progressed 20 per cent faster than those with less daylight and that classrooms with skylights were associated with a 20 per cent faster rate of improvement.[9]

If designed skilfully, a building will help to make children's experience of their early-years care a secure yet varied one. Defining value is a discipline which goes well beyond the initial capital cost, which are sadly the traditional parameters which define quality in public architecture. Today as much as the children, the nursery environment needs tender love and care throughout the years of its use.

Although nurseries and children's centres take different forms in the United Kingdom, to suit a variety of local conditions and needs, many include daycare with associated family areas such as therapy, training (for work) and counselling rooms to provide wider social support. They offer quality-assured family services to some of the most disadvantaged people in England, many of whom are perhaps using childcare for the first time. This is now happening within purpose-made buildings. However, generally speaking this has been a standard in most European societies for generations, undoubtedly a defining aspect of the more generally community-minded European citizen.

Many of these new UK facilities have not necessarily benefited from big budgets, yet they have, by and large, provided good adaptable accommodation; they are proving to be facilities which nurture the 'home from home' principle for those using them. By 2010 the ambition is that there will be around 3,500 children's centres, one within walking distance of every community in England. In this section we describe a different range of children's centres, from high-budget purpose-designed facilities to premises which have been adapted on shoe-string budgets. But first of all, how do we define what is a children's centre?

A children's centre is whatever it needs to be in order to meet the needs of young, vulnerable people, usually but not always within the most deprived situations. However, it is blindingly obvious that the route to a happy child is the ability of his or her parent(s) to provide good responsible care, love and nurturing within the home. This means that parental needs must, if possible, be met, as well as those direct needs of the child for secure care and education in nursery. The children's centre is not necessarily just for children. At it's best, the children's centre is a community facility where the whole family can visit to access a range of integrated services provided by the Local Authority. Clearly this recognizes the HighScope experience (and the nature of most European childcare systems), that parental education and support is as important for the well-being of young children as care and education is for the young children. While childcare will almost always be the main function, there are some children's centres which are almost entirely devoted to support for parents with little or no childcare on the premises. Childcare/education may in these circumstances be undertaken in the form of satellite children's services located within the community, usually in the homes of child-minders.

For example, the Familio Children's Centre in Mansfield, Nottinghamshire, is just such a centre. Mansfield is a former mining town in the north of the county, which suffers from high levels of ingrained unemployment which now goes down through several generations since the closure of the mining pits in the 1980s. Along with unemployment comes other byproducts such as drug use, poor health – both mental and physical – and the highest levels of teenage pregnancy in Europe. Many new parents feel isolated in what traditionally (only 20 years ago) were tight, closely knit urban communities with the traditional employment situation adhering to the male-breadwinner model, with the man employed in the pit. Some parents are so young they are by certain definitions still children themselves. The aspirations of juveniles are so low and unemployment is so institutionalized that when we asked many young teenagers if they would be prepared to leave Mansfield to find work, the answer was a resounding 'no'.

The new Familio Children's Centre feels slightly detached from the community it serves, as it is situated on an edge-of-town site at the rear of a primary school at the end of a residential street. The logistics of this are as a result of a slightly unfortunate last-minute government decision to locate new family centres on school sites. Previously, sites in town centres close to health centres and shopping centres had been preferred. Perhaps a lesson for government here would be to avoid centralized policy making that discounts the particular needs of the local community. Particularly frustrating to the users is that very little local consultation took place. Nevertheless, during its brief period of operation when we visited, the new centre has become an essential focus for family life within this impoverished community.

The new building was opened in September 2006. However, it is important to point out that a range of support was already in place, in the form of services which were taking place in family homes by way of visiting social workers operating from their 'car boots', and in existing community centres and church halls within a ten-mile radius of the new building. What the new facility has done is perhaps less to do with childcare, and more to do with parental needs; the new building provides a hub where all team members and parents can come together in a safe and secure environment. There they can engage with others in their community, thus restoring a fundamental social service, the ability to meet and form networks in what might otherwise become an isolated and isolating existence. It is a children's centre, but childcare is only one small element of the services it supports. The childcare takes place side-by-side with parents who are being trained for employment. The centre has 30 staff providing a range of outreach services, some of which takes place on site, with a good proportion taking place in other facilities, including home visits and part-time sessions in church and community halls. Here we see 'the hub and spoke' concept promoted by the government, with over 50 per cent of the accommodation relating to staff supporting and providing services on or outside the premises. In this particular context, Familio is not an unusual model for the modern children's centre located in socially dysfunctional conditions such as this, where generational unemployment is a key factor in community break-down.

Welcoming and easy to access, these places bring together limited but flexible care for children, early-years education, family health provision and other support services. Help will also be available to parents who would like further training or for those seeking employment. In these situations, safety and security are of primary importance, and while the architecture is rarely ground-breaking or iconic from a design perspective, mostly a simple functionality is the primary expression of care and concern to communities which have missed out on the economic growth seen in other parts of the United Kingdom over the past decade.

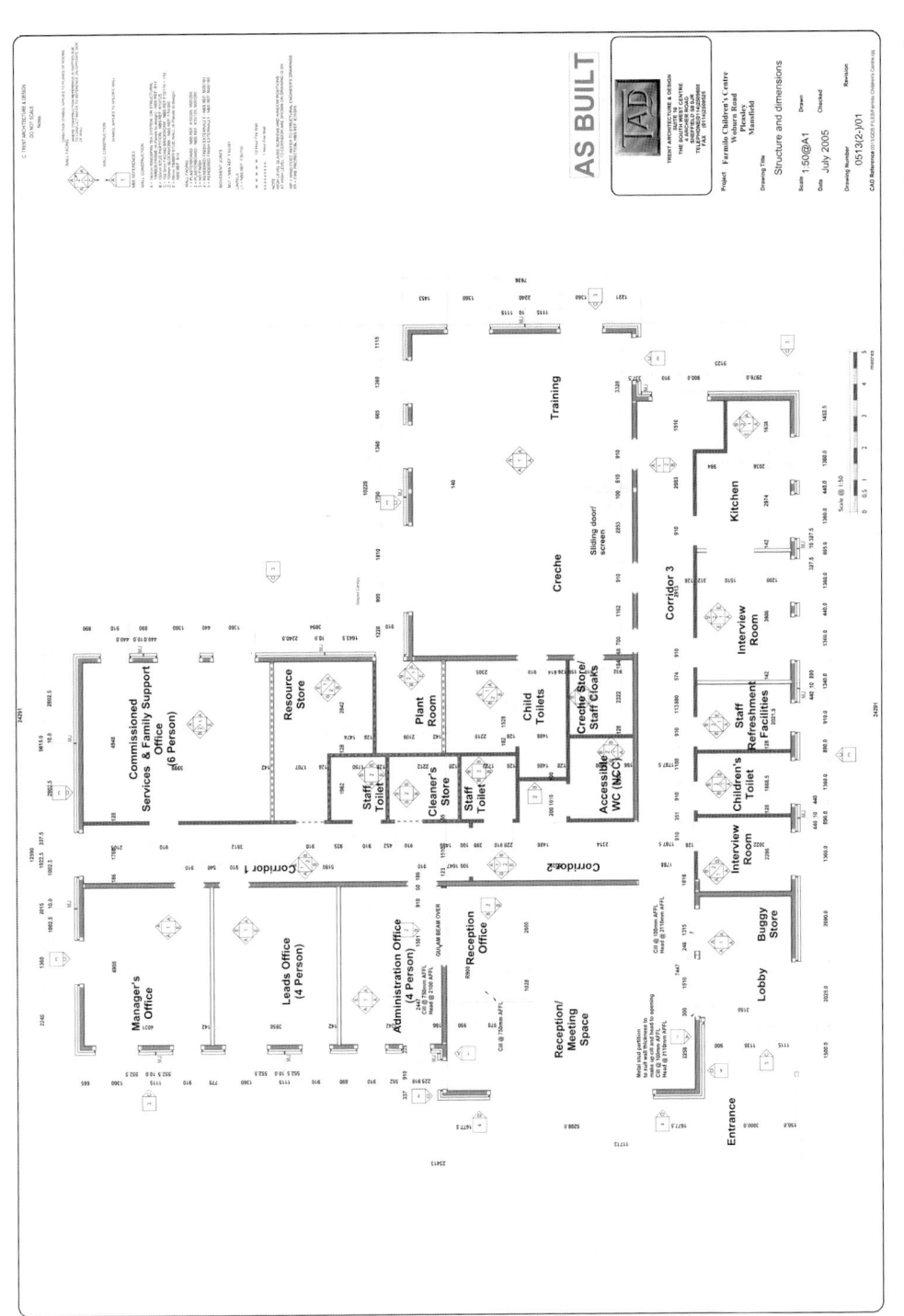

The Familio Children's Centre in Mansfield, Nottinghamshire. This is a deprived community and the majority of support to young parents is in the form of home visits and other so-called satellite services which take place in church halls and community centres closer to home – this is the hub of the spokes. The schedule of accommodation is dominated not by children's spaces but offices and other adult-oriented services intended to organize the distributed sessions. Designed by Trent Architecture & Design (2005).

The sustainable nursery

The entrance lobby reception desk has a child-height toy shelf to distract children while their parents are busy talking to staff.

If the mark of a successful intervention is the degree to which gratitude is forthcoming from its users, Familio is very successful indeed. Parents are simply happy that they have been considered and given something which is specifically for them. It is still a children's centre, but the emphasis in this particular location is more on supporting parents and less directly on the provision of high-quality architecture for childcare. It may not be a particularly inspiring building, but then most family centres are a compromise balancing

Outside everything is functional and oriented towards adults rather than children, with car-parking at the front for staff to drive out to the satellite services.

tight financial deadlines and stringent budgets. Small design wins usually come about as a result of immense user pressure combined with an architect willing to go the extra mile in terms of user consultation and value for money in design. At Familio, a simple example of this is the wide reception desk, with its adult-height work-top and child-height shelves running along the front, which are filled with little toys. This is welcoming to both parents and children – often, isolated people who may be reluctant to come to such a public place are 'nudged' by such simple architectural gestures.

At Tolladine Sure Start and Community Centre a range of community facilities are brought together in a single, integrated complex right at the heart of the residential community it is intended to serve. The project comprises part new build extension and part major refurbishment of a low-grade 1930s church building, more recently used as a community centre. Working with community groups, the architects have integrated a 26-place crèche with parent training, a new entrance and rejuvenated foyer space by the judicious coming together of old and new. There is a dedicated youth venue, a small chapel, offices and other shared ancillary accommodation retained and refurbished from the original dilapidated buildings. The new facilities wrap around the existing structure to give a faintly confusing image of a funky, new, metal-clad night-club on the one hand and a stodgy old red brick structure on the other, garden side. On the inside, however, there is no such ambiguity; it appears to be a completely new and vibrant sequence of linked and complimentary spaces.

The key architect's concern was how to address the architectural tensions between a multiplicity of different users and the range of architectural responses suggested by this diversity, implying the need for anonymous non-descript, so-called flexible space planning, and the specific spatial requirements of young children, who were to be the prime user group, certainly during the daytime hours of its usage. The designers developed a concept they called 'own space shared space' in which small parts of the accommodation could be marked or personalized by users themselves to provide distinct areas of dedicated architectural space within the overall framework of the communal whole. These interventions enable each child and parent or parent-to-be to find a territory of their own within the new building.

The building is divided hierarchically into a range of different scales, from the spacious foyer, to the niche, nooks and recesses providing places for a broad spectrum of moods and functions. There are places for group play and community activities, there are spaces for small-group play or for more personal one-on-one interaction. There are even corners which are described as 'withdrawing spaces', specifically for children to find their own place to play alone if they so wish, to be quiet and reflect, insulated from the hubbub – an unusual allowance indeed.

This spatial hierarchy is also expressed in three-dimensional form; for example with the design of the main crèche space which is tall, light and airy so that it can accommodate 26 children with their carers at one time. There are subsidiary spaces of diverse scales, ranging from a large bay window with long, low seats to narrow recesses which can be adopted as dens for small children. These child-scaled nooks offer quiet spaces set apart from the main activity areas. While it may not suit carers who demand

The sustainable nursery

Tolladine Sure Start and Community Centre, the niche seat with storage under, designed by Meadowcroft Griffin Architects (2006).

The child-height window and the tree seat outside forms a close relationship within the imagination for children using this centre in Richmond, Surrey. Designed by Mark Dudek Architects (2004).

constant supervision of the children, this shift in scale and enclosure is a valuable theme in early years (see the example of Lustenau, below). It is clearly intended to encourage exploration of identity through play, with architecture providing imaginative possibilities for children to do this readily.

According to architect Phil Meadowcroft, play is the most serious form of activity for a young child; it is a way of discovering self and constructing relationships with the wider world. Finishes are intended to encourage unfettered play in all its complicated manifestations, so messy play is enabled by way of wipe-down surfaces and child-height paint sinks; there is direct and easy access to the storage of resources to facilitate quick and easy access to children. Because of the multiple usage of the centre, these resources must be secured at the end of each day, which is another challenge to designers in this area of design. The sensory qualities are also considered to be important so, for example, existing rough-hewn brick walls helps create textural variety within the new building, all at child height. Shaded recesses bring a strong sense of nature and the outside world to the interiors, allowing it to change with the seasons and to reflect sunlight off a spectrum of colours and textures, from rich red brick to smooth luminous yellow plasterwork and aromatic cedar cladding.

Externally, the spaces form a new front entrance to the existing building, strengthening its local identity and giving it a civic presence which it simply did not have before, helping to foster a positive sense of community. Outside, the new crèche space is clad in expanded metal panels designed to act as a trellis for evergreen and seasonal flowering plants, which forms a protective hedge, preventing children from climbing onto the roof. The entrance and youth wing have a more open treatment leading into a top-lit foyer space which provides a welcoming route into all areas of the centre. This space between the new and the existing provides a generous zone for a new community café and meeting space, transforming what was previously a left-over alleyway area into an important public space. Overall it is an extremely economical new building which integrates old and new structures into an attractive centre that is welcoming for the entire family. This approach provides a neat reference back to the original building, without compromising the functional or stylistic qualities of the new architecture.

A similar scenario faced the author as the architect for a part-new and part-refurbished building, Windham Nursery and Croft Centre for children with learning difficulties. Here the situation was that two very old buildings undertaking childcare and support, albeit in different forms, were divided by a high brick wall and confined within a tight site. We suggested that the two institutions collaborated in their common aim, namely to support the welfare of children in the community, partly by sharing facilities such as the sensory room and the community meeting rooms, and partly by using the space between the two existing buildings (where the wall was) to provide a new entrance and community meeting room. A further merit of the site was the confined urban park behind the site, which was effectively no more than an occasional evening hang-out for local teenagers and dog walkers; the existing buildings turned their backs on what could become a much more child-friendly environment.

Because it was very early in the whole process of early-years provision in the United Kingdom, problems of confidence among the childcare

The sustainable nursery

workers militated against the possibility of much experimentation, particularly in relation to trying to amalgamate the two sides of the site. However, removing the wall and having a joint entrance was eventually agreed, and the space between the two existing buildings provided a frame for the new entrance to form a community space beyond, which looked over and accessed the park at the rear. The new and the old come together in a curious amalgamation of left-over doors and windows, which creates an interesting juxtaposition of light and shadow which makes much more of the facility than was previously the case. Ten years after it first opened, the Windham Centre is currently undergoing a part re-build of the Croft Centre element (which originally was only lightly refurbished). However, the basic structure established by these original bold moves has enabled the whole to become much more than its former dislocated parts. Furthermore, the landscape has now matured, and the interiors have been adapted by the users through the introduction of new furniture and decorations. It is a treasured part of local community life, a minimal investment having a positive effect on generations of young children.

A more substantially funded project which consequently has much higher architectural aspirations is the Effra Early Years Centre in Brixton, London. This new building is located in a quiet residential street. There is ample parking for drop-off on the street, but it has no dedicated parking. However, most parents appeared to be arriving on foot when we visited, a much healthier way to go – clear evidence of the transformational nature of the facility. The building is not immediately apparent as it is set back from the street edge; however, a generously proportioned entrance gate in slatted timber set into the high brick perimeter wall provides views of the building

The tree becomes an attractive feature with the addition of this child–adult seat, an essential part of the 'second layer' which makes the facility truly child-oriented. Mark Dudek Architects, Windham Nursery and Croft Centre.

from the pavement edge without compromising on security. The first sight of the building through the entrance gate provides views of a distinctive blockish building in three sections: the nursery to the left, the community hall to the right and the centrally positioned two-storey entrance block. Either side of the entrance route are children's play areas, which are rather hard and austere at present, but each has its own canopy for protected outside play.

The façade is in highly coloured hues in the yellow to orange spectrum. The central entrance block has playful signature lettering high up, which is also visible from the street entrance. This combination of bright, advancing colours and good graphics gives the building a strong civic presence, almost on the level of product branding; it is discreet yet vivid at the same time. The confident use of colour applied in a harmonious way is particularly valuable in making more of the simple architectural forms.

When tight budgets restrict the use of expressive natural materials, colour used in this way is extremely effective in establishing a positive image and giving a strong impact within the wider community. In my view far too few children's buildings exhibit this assured use of external colour. Colour seems to be a dimension which many architects steer clear of at this level; often one suspects this is because they do not have enough confidence to express much more than the building's basic form and function. Effra has important lessons on this level for other architects developing urban buildings of this type. It is important to emphasize the strong civic presence the building has within its community, provided by its decisive architectural stance.

Internally the building reflects a similar clear organization, albeit with a more restrained use of colour. The planning strategy positions the entrance and reception area centrally in a double-height space, which according to the architects is like a tower. The double-height volume gives it a sense of spatial hierarchy, a public face to a private facility. The windows to the main children's activity areas along the south wall opens out under a 4 m deep canopy which has aluminium shutters at its edge, allowing children to store their toys and bicycles securely overnight. During the day the shutters are rolled up, giving children a large open-air play space – a natural extension for the internal activity areas. The north corner of the complex contains staffrooms and other related facilities, and on the east there are meeting and training rooms, WCs and a big community space that can be divided in two with electronically operated folding and sliding doors across the entire width of the room. With its separate entrance which can be used when the rest of the facility is closed off, and flexible multi-functional plan it is a community space in the best sense, used by diverse groups in the evenings and at weekends, as well as providing crèche space during the days.

The main nursery activity space, which provides up to 90 childcare places, is a single open-plan area, which is unusual for a children's centre where the enclosed smaller group home base concept predominates. Integrated wall storage units and an extended canopy to the outside provide flexible and accessible storage as well as practical all-weather play. However, the acoustic performance of the main internal activity play space is a real issue. With its open form the noise from 90 children playing in the same space is at times extremely uncomfortable according to some staff we spoke to. There is perhaps more of an aesthetic architectural agenda operating at the expense of the user's comfort in this regard. While it may look pretty, the 'cloud' ceiling does not particularly help, although it does look good in the architect's photos. However, Effra is on the whole a good and valued family centre, not least because children's features, such as the spacious and social children's toilet areas, have been designed with the child in mind with, for example, a bank of hand-wash basins facing each other for enhanced social interaction, and most importantly, good natural light and ventilation with a view onto the garden. It is clear that the architects have understood the fundamental principle that, for children, every space is a learning space.

A kindergarten in Lustenau has a similar coherent image to Effra, with the use of high, arched volumes for the activity areas and a section which is cut into the existing site slope (see page 78). There is an integrated slide which carries the theme of play into the actual image expressed by the architecture to the surrounding community. It has a very distinctive presence and it looks and feels like a public building in its own right, as it should. However, unlike Effra, the children's spaces are subdivided into distinct age-related groupings which prove to be more acoustically manageable than Effra. The cross-section illustrates distinct volumes within these play areas. First, the low, intimate children's activity spaces together with ancillary services such as toilets and storage at the rear. Second, the main volumes overlooking the children's garden with a floor-to-ceiling height of 4.2 m and automated up-and-over door/windows opening the entire interior to the outside on sunny days. Thus children can enjoy clear access to the garden outside – there is no distinction between inside and outside. According to the design architect, Konrad Frey, the building adopts clear child-oriented themes:

> My first thought was play-caves in the slope. And to show the children a different world to the one familiar to them from their homes. To recover the built-on land on top by folding up the countryside and take the landscape inside. One can sense the rolling hills from inside.... We have spent as much time on the issue of acoustics as anything else, when the roof on the south façade is extended then it becomes an open air classroom. The transition is significant: from raw acoustic plaster via a textile membrane to the translucence of the polyester shell which juts out above, and then comes the sky. The building itself as a toy.[10]

The green nursery

Increasingly, there is evidence that the predictions of global warming and extreme weather conditions are becoming a reality. The scientific community is almost in unanimous agreement that these changes to the climate are a direct consequence of man made pollution and land use alteration. A recent Royal Commission report made strong recommendations to both improve energy efficiency and increase the uptake of renewable energy sources, if we are to control the environmental impact.[11]

As I write, the second and third waves of children's centre construction in England are underway. It is clear that lavish funding is no longer on the agenda. However, good child-oriented environments can still be created on relatively modest budgets. For example, Discovering Kids Playgroup was established some ten years ago in the rural village of Loup, County Megherafelt, 40 miles south of Belfast city. Clare Devlin, a working mother living locally recognized the value of childcare provision, not just for her own children, but also for the benefit of wider community cohesion, providing a non-sectarian facility in the immediate aftermath of 'The Troubles'. In 2002, along with a group of like-minded mothers, they established a 15-place sessional nursery in a disused building owned by the church. The building itself was an old nineteenth-century school hall, which despite its rural location, disappointingly had no direct access and no views for children to the outside areas with its lush fields and stormy winter skies beyond. To get out, children had to use a narrow access alley leading from the main entrance to the outside spaces around. Once outside, children were faced with the daunting prospect of a hard asphalt yard surrounded by high brick walls; the only view was onto the adjacent cemetery.

In 2003 the childcare committee recruited specialist childcare architects Mark Dudek Architects to turn the inaccessible storage yard into a new children's play garden. Partly supported by a grant from the EU Peace and Reconciliation Fund, limited funding to the tune of only £60,000 has enabled the creation of direct access for children to the outside areas by way of a new window door from the play area to the yard, together with a covered outside canopy complete with sand-pit so that a sense of the outside environment is available to children even during rainy weather. More importantly, the paucity of funds required that a certain amount of recycling, both in terms of materials and people, would be required. As a result of a determination to get gift labour and other free materials into the building process, the entire outside yard area has been made child-friendly by way of a number of features such as a cultivation garden, a raised deck running along the side of the yard, new climbing equipment and generally a complete softening of the surroundings through the addition of planting and interactive wall panels. What was previously a hard asphalt storage yard has been transformed into a vivid and highly attractive space for imaginative play and learning.

Although modest by the standards of Effra and Lustenau, the project has completely transformed the experience of play and discovery for children and their parents, mainly because it has been a community-up procurement strategy rather than a government top-down process, where the enthusiasm and commitment of the users has been harnessed to the benefit of the architecture. Working within a tight budget, the project has benefited from gift labour and extensive community support throughout its construction. It is neither sophisticated or completely finished, but there is something extremely appropriate about it within this divided political context. For example, while the new ceilings in the activity areas are not particularly stylish or sophisticated, it is surely wrong to be critical from a design perspective since they were supplied and fitted free of charge, as a gift to the community from a local supplier. A number of similar 'gifts' have enabled the full scheme to be completed.

The sustainable nursery

The church hall at Loup, County Megherafelt showing 'before' and 'after' images which illustrate the stark contrast between an environment which is inappropriate for children and one which is ready for play. This conversion/extension, by comparison to the hi-tech, high-budget Lustenau Kindergarten, was built on a shoe-string budget using gift labour and a €100,000 grant from the EU Peace and Reconciliation Fund in Ireland. Designed and supported by Mark Dudek Architects (2006).

The plan illustrates six basic steps to raise the quality of the existing building to make it suitable for use by young children:
1   A new opening is formed in the existing building; this facilitates run-in–run-out play.
2   A new, secure, covered canopy and sand-pit have been built; this extends the field of learning as doors can be open even when it rains.
3   A new raised play deck has been built along the boundary to the adjacent church-yard; this provides space for children to 'escape' from gross motor play and engage in more intimate fine-focus activities.
4   A child-safe ramp has been formed to connect the lower level to the new upper-level sand-pit and cultivation garden; this enables children to view the surrounding countryside and engage with soil and sand for tactile play.
5   New planting around the perimeter softens masonry walls; more cultivation provides a rustic softening quality which brings high brick walls down to the scale of the children.
6   Relocation of the existing storage shed and creation of a new secure entrance; this uses existing equipment to ensure a sense of enclosure and enhanced security to what would otherwise be open to strangers.

The project illustrates that with little funding a children's environment can be completely transformed for the better on a tight budget. Strong, cooperative communities can make a little go an awfully long way. Normally, the sustainability agenda is all about generating electricity from the sun or sticking a green roof on top. Here, it is much more to do with a state of mind, a proactive response to the will of the collective community. The 'big society' at work, perhaps. The Loup project has recently been adopted by the Rural Development Council in Northern Ireland as a model of best practice and it prospers as subsequent parents have their own offerings to make towards its overall environmental and spiritual quality.

A similar low-cost project that also provides good-quality daycare is the Burley Children's Centre in a suburb of Leeds. Located in a converted infant school building that has had a minimal amount spent on it over the past ten years, the building nevertheless provides an excellent environment which is user friendly, atmospheric, full of light and child-oriented features which continues to welcome new generations of parents and children. Yet it has a distinctly 'DIY' quality.

Walking around the premises accompanying the head of centre, Mandy Quayle, is an enjoyable yet slightly disconcerting experience. It is pleasant because children are clearly happy and well looked after, and superficially the building seems attractive, child-centred, comfortable and spacious with lots of special areas decorated to a high standard, albeit in a slightly Heath Robinson manner. However, it is disconcerting from a professional perspective because this adapted environment does not comply with what might be viewed as widely agreed childcare environmental standards in design. As a cynical comment, how does an architect actually earn a fee by working with this particular childcare community?

For example, the official guidance for any new childcare facility worth its salt will demand a level threshold between the inside and the outside. It stands to reason that children running outside to the play garden

The story corner at Burley Children's Centre, aesthetically pleasing at a fraction of the cost of an architect-designed version; it is a home-made space which has tremendous appeal.

The sustainable nursery

should not have to negotiate trip hazards. In addition, there is access for disabled people to consider (a statutory requirement for any new building), which at several points simply is not complied with. Similarly, the Burley entrance area has hardly any civic presence at all; it is a long, featureless access route down one side of the building with a plethora of hand-made signs to welcome parents and provide guidance on how to enter the building. There is some planting, and crucially a large mature oak tree which spreads itself across the threshold like nature's canopy, which goes some way towards softening the entrance zone. Yet beyond this, the entrance hardly announces itself to the outside world at all. It is a long way from the high-profile presence Effra has with its sophisticated large-scale graphics and brightly coloured façades, loudly exclaiming its presence within the community. Yet Burley emanates such a positive ethos that it does not seem to matter. It is worth reflecting on the special qualities which make it so.

One of the keys to Burley's success is undoubtedly the partly self-made nature of the environment, a process which has been largely facilitated by the staff, which might not have happened if the users had been given a brand-new building. As a practising architect I say this with some regret: often high-end architecture design inhibits the users; they simply find it hard to relate, perhaps believing that to make any changes to a given environment goes beyond their pay grade. The building is too finished, too perfect and complete to the point of unassailability. They may look back to their old building with some nostalgia, as a place in which they felt relaxed enough to paint the doors or build a cupboard themselves. Their sense of empowerment made the place fizz with energy and good childcare practice. It is a dilemma. The new build is often 'too finished' and over designed.

Obviously an architect has different priorities to an early-years teacher and, conversely, it never ceases to surprise and disappoint me as to how much staff can subvert spatial quality within a new building by overelaborate decoration. While I am fully aware of the architect's predilection for often obsessive spatial clarity – sometimes referred to as minimalism – this tendency to DIY decoration can, in the wrong hands, totally obliterate the architecture itself, thereby negating the possibility that a well-designed environment supports children in the area of their own spatial awareness and in ways the architect and his team had originally intended. This has been a long-held personal view; however, at Burley this is certainly not the case. Rather, the decoration is complementary and created with a keen artistic eye which in its totality helps to make this an attractive homely environment conducive to child development on many levels.

It is important to recognize the strong leadership and stable workforce that has enabled this to happen. Part of the problem for any childcare business is the sense of transience (relating to inherently poor pay and insecure career prospects, perhaps), which makes people change jobs frequently and move on. When a successful team is broken up, children will lose the continuity of a carer with whom they are emotionally connected, and in the case of some insecure children, individual staff members who are even acting as surrogate parents; children may become more anxious as a result, which will inevitably undermine the ethos of the group. Equally, the understanding of longstanding staff about the needs of the environment and their sense of being an essential component of the system and part of an ever-improving environment, one which evolves over time, will be totally undermined also. It is somehow as if the very act of seeing and working with the spaces on a daily basis, year on year, turns the staff themselves into designers and improvers of the building even if it is on seemingly superficial levels, a feeling which comes along only when they have their own sense of stability and belonging to the particular environment in which they work and feel safe.

At Burley the staff are a longstanding effective community in their own right. They have collectively shaped and moulded their environment over a number of years to form a place where they all feel comfortable, both individually and collectively. To reiterate, in some senses it is a better outcome than might have been the case if a completely new state-of-the-art building had been imposed upon them (to use a deliberately pejorative term) with all the disruption that would inevitably have entailed. This is the type of ambiguity we experience with much early-years architecture in the United Kingdom at present. Sometimes modest and low-cost yet skilful interventions tailored specifically to their ongoing needs and made by the users themselves has more value than lavish new builds conceived at a distance from those who are expected to use them. This of course is also a factor that relates to how well the design team consulted the users in the first place.

At Burley they are very aware that they must work within tight budgetary constraints, yet they also have a clear vision as to where they wish to be with the design and development of their centre in 5–10 years' time. Aided by relatively small periodic tranches of funding from the Local Authority, the team have initiated a range of gradual improvements (with the help of a local building contractor whose children originally attended); for example, they have infilled and extended an existing open courtyard between the childcare and the community spaces with a simple, flat, glazed roof which creates a welcoming and secure entrance. Enclosing the entrance has also created an additional enclosed area between the adult care wing and the nursery itself. Again, this pocket of captured space has been worked over with a new roof and extended heating system. Thus it is transformed into an internal area which is now used as a children's gym and chill-out zone, providing an extra dimension to the childcare curriculum. It is an immensely valuable new space which arguably would not have been affordable in a conventional new-build scheme. They have also renewed and improved the children's toilets to make a pleasant, usable space right at the heart of activities, and developed an open-plan kitchen area to provide good-quality food which promotes a culture of domesticity. They are currently developing what they describe as their 'magic garden', a small but imaginatively designed area at the back of the main activity areas for exploration and focused outdoor play.

Externally, the building is incoherent and far from attractive, yet the most abiding feeling of this nursery is of the spacious and friendly interior, with the right number of children for the space, and enough nooks and crannies to allow children the feeling of escape from the supervision of adults and an interior design which has had a lot of thought lavished upon its improvements. Clearly the improvements are gradual. Consequently, it is a non-institutional environment which neither intimidates or patronizes its users; the very fact the staff have had the time and space to dream about the various improvements which have accrued over time gives it that unique quality. In answer to our own rhetorical question of who pays the architect's fee, no-one does, as no architect has been involved with the project.

However, it has to be said that the downside of Burley is the issue of environmental comfort and from an ethical perspective, the green sustainability agenda in buildings is now widely accepted as being an important countermeasure to global warming. In the United Kingdom all new buildings must now comply with a rigorous insulation standard which is a statutory requirement enshrined in Part L of the UK Building Regulations. Since Burley is an old building it does not in theory have to comply with contemporary building regulations. The windows are still the original single-glazed leaky design and the heating is a somewhat anachronistic blown hot-air system (at head-height), which is by all accounts extremely inefficient and costly to run. Although it was not evident when we visited over a couple of cool overcast days during the spring, it is likely that during the long, hot summers to come, it will be an extremely uncomfortable environment with a tendency to overheat. Equally, on the 16 or so coldest days of the year it is going to be cold and draughty (although this was not a problem reported to us by the users, who stated that children tend to keep their coats on during very cold days since they are in any event outside for much of the day). However, to further muddy the argument, often so-called state-of-the-art new builds also get environmental comfort disastrously wrong, so that during some extreme weather conditions the traditional leaky single-glazed facility seems no worse than some of the modern multi-million-pound investments.

An example of the above is the Lavender Children's Centre in South London. A well-funded new build and part of the government's original Sure Start programme, it provides a range of facilities including 100-place daycare units with a large, secure play garden at the rear. It was designed as a pre-fabricated modular construction, with passive ventilation throughout. Unfortunately, during some hot summer days this building is extremely uncomfortable. Given the laudable aspiration to make an economical pilot project with sustainability a key priority, it is important to flag up the profound difficulties the users are currently experiencing, particularly in relation to environmental comfort.

When we visited, the following comments and issues were raised by some parents, but mainly by the receptionist/administrator. They are, I think, broad and wide-ranging but of interest:

- too hot all year round (this from all of those questioned) – they do not see air conditioning as a solution to the problem since it restricts easy run-in–run-out play;
- window openings are restricted and limit airflow – Debbi suggested that window stoppers would prevent children from trapping their fingers, which is a present hazard; however, the main problem is that they (the windows) are large and south-facing, with restricted opening;
- not enough storage space;
- not enough definition to the spaces (colour contrast) for visually impaired people who need clearly defined contours or coloured walls and doors to stop them bumping into things;

The sustainable nursery

- some said they would have preferred there to be more colour in the spaces for purely aesthetic reasons;
- floor-to-ceiling glass is a problem for children who do not realize it is glazed, as they may run into it! They have put stickers on the panes, for clarity;
- they do not like coloured glass as it is too strong and casts a shadow; they seem to have a particular aversion to the yellow glazing used in a room with a computer; one employee claims that it (the colour yellow) makes her feel unwell;
- floor-to-ceiling doors in the milk kitchen had no glazing panes and kids got bashed on the other side of the door because they could not see others, so they cut them in half (make them into stable doors was one proposed solution);
- tarmac areas in playground subsided (problem now rectified);
- some said the lighting is perfect, others said the lighting is too bright;
- would have liked an even larger cooker;
- staffroom is too small; there are often up to eight staff members at one time;
- ceiling hatch (roof light) in the staffroom is broken and very hot as a result;
- water flow in the staff kitchen and loos is very high pressure, so water tends to spill over onto loo floors and kitchen surfaces;
- sinks in loos are too small;
- it is a good, secure building;
- there is a great layout generally.

Although users of new buildings are generally positive, in this particular situation they all raised environmental comfort as the major concern during the interviews. Indeed, it is such a problem that air conditioning has now been installed, at additional cost and as the comments point out, seemingly undermining the architects' primary concept that children should be able to freely run in and run out between the activity areas and the garden in the summer as the doors could be propped open; with air conditioning you can't leave the doors open, which makes it very difficult for young children. The original intention to create a sustainable building which was passively ventilated and illustrated its sustainability credentials through good design has become the very problem which prevents this from functioning effectively, despite having design expertise from sustainable design consultants.

I mention this project to highlight the pitfalls that some designers open themselves up to when they design nurseries. The priorities in early years are not the same as when designing almost any other building type. In this particular situation, the lightweight timber fabric used in the construction matched with low ceilings and large south-facing windows has made this building problematic and suggests that insufficient consultation took place between the design team and the users. It also hints at the responsibility users have to understand what they are getting and to emphasize critically important aspects of the functionality agenda before construction. Architects and developers must take the views of the existing community into account in relation to the existing environment (if there is one), to recognize its positive aspects in

The sustainable nursery

use and match realistic aspirations to the budget available. Architects should never view an early-years project as a potential prize winner; rather, they should take priorities of its users on as a sort of family, to be supported and understood in a spirit of compromise and love.

Today, sustainability is the key ingredient in high-quality contemporary design, not just because of the threat of global warming, but also because it is important to convey the importance of sustainability to young children so that the building itself becomes something of an ethical teaching tool. Contemporary standards of heating performance and efficient building fabric systems means that ventilation and cooling may be the key dimension for any new childcare building. For example, the author has recently completed an early-years building that was constructed from pre-fabricated, factory-made timber panels. Brought to site on the back of a lorry, it is literally bolted together in a matter of hours. It is an extremely efficient procurement route, minimizing on-site construction time and providing a dry interior within which other workers can operate soon after its arrival on-site. Not only does the manufacturing process provide a highly insulated system because the panels fit together so tightly, it also fulfils the important modern criteria for building design to minimize air leakage as much as possible. It stands to reason that the less cold air escapes during winter, the less heat you need to provide. However, it is then critical that in the summer the building does not become too stuffy, so air conditioning is required. We ensured that this would not be the case by creating a high, sloping ceiling which acts as a sort of plenum, forcing air in at the low north-facing end, which then rises naturally to the top of the area, where it is expelled through wide high-level windows.

It is also highly efficient in terms of heat exchange, with a comforting underfloor heating system throughout running off a bio-mass boiler, which is powered by recycled timber pellets and an underground heat exchange system which utilizes the natural cooling function of the earth in the summer and generates heat during the winter. This simple combination of high thermal efficiency and effective passive ventilation provides remarkably low running costs, which are complemented by a solar heating panel system located on the roof to provide additional free power to heat the hot water services. The form itself aids ventilation with a profile which promotes through-ventilation from low to high, so called passive ventilation as opposed to resorting to mechanical air conditioning. It is an extremely comfortable and efficient environment which promotes sustainability to all of its users.

This building is the first stage of a longer-term development which will see the entire Stanley Infant and Nursery School, South West London rebuilt. The circular form functions well in terms of the curriculum, giving it lots of wall space and facilitating a flexible floor plan in which staff can provide a diverse range of activity zones. However, it is also a statement in its own right. Seen within the context of the existing decrepit school buildings, which are rectilinear concrete blocks with too much glazing for comfort, it is an unusual building in that glazing is reduced to the minimum without having to resort to artificial lighting during the daytimes, the theory being that for young children, a large proportion of their time should be spent outside in the garden.

Its architectural novelty is like a breath of fresh air that makes parents sit up and children inquisitive. This is especially the case for older children from the main school, who do not actually benefit from using the new building themselves. The end result is like an expression of what is possible on minimal budgets; it shows the power of architecture and design to communicate important messages about sustainability and childcare services. Early-years architecture is almost uniquely able to use these expressive playful forms without necessarily jeopardizing functionality or image. A school seems to be an altogether more serious place and does not always justify this lightness of touch that is permissible in early-years buildings, with their more playful message. Aside from all the sustainability gizmos that enable children to see and experience the ways in which power generation is happening, its shape gives them an instant child-like perception of space, a message of home, perhaps.

There are a number of examples of this type of progressive early-years architecture that are worth describing. The common bond these examples illustrate is an imaginative engagement with childhood perceptions, together with sustainable design that is ahead of the curve, all expressed through the architecture itself. For example, the Paradise Park Children's Centre in Islington, North London is a long rectangular building running parallel to the street, which provides a full early-years service to local people. On approaching from the main road, the feature which strikes you immediately is the main façade, a wall which is covered in planting, a little like a meadow growing perpendicular to the ground. It features an irrigation system which ensures that the growing medium is consistent throughout the year. It is a powerful, almost magical statement which shows that this is a special building, and one which the designers have taken great care over. Stimulating and engaging to children, it is a statement promoted through innovative architecture of the highest quality which shows the symbolic affinity children have to nature. It never ceases to give pleasure to users and passers-by alike.

The Paradise Park Children's Centre also works as a crucial example of sustainability in action with the use of natural or recycled materials throughout, low-energy heating and lighting and what the architects describe as a 'brown roof' on top. This entails the re-use of excavated materials from the foundation dig-out transplanted onto the new weatherproofed flat roofs. What is important is that rather than being carted away to distant landfill sites, the spoil takes the short journey from ground to roof. This provides a heavyweight and therefore highly insulated internal environment which responds slowly to the changing solar conditions, absorbing the radiant heat of the sun on the hottest summer days, moderating temperature without recourse to mechanical cooling. With its dense thermal mass acting as a highly insulated warming blanket during the cold winter months, it is proving to be a highly efficient system with significantly lower fuel bills which over time will offset the higher cost of the build than might have been expected from a lightweight form of construction. The architects believe that this brown roof is far more sustainable and a genuinely eco-friendly strategy, since the waste used on the roof is of the locality rather than being imported greenery from a distant source. It is envisaged that the natural seeded planting which will gradually emerge from the 'brown' waste will produce the same indigenous species of vegetation, insects and bird life to be found locally. It is in effect like pulling the ground up and over the building, thus completing the greening of a truly sustainable nursery.

However, buyer beware: since I wrote the previous two paragraphs, significant problems have developed in relation to the 'green wall'. It is a salutary tale, yet one which holds important lessons. Due, one suspects to the lack of maintenance of the irrigation system, Paradise Park Children's Centre now has a brown bolding wall presenting itself to the public street. We visited in the winter and the effect is sad and depressing, an essential component of the green sustainability message which has quite simply gone wrong. Maintenance is always an issue, particularly where technically innovative solutions are proposed.

Sometimes architects can be too experimental. The early stages of a 'green wall' featured on this nursery in North London illustrate the feature at its best; unfortunately, three years later the complex irrigation system had failed, most of the planting had died and the building appeared sad and neglected. This certainly illustrates the need for good maintenance, which is not always possible during times of budgetary cost cutting by local authorities. Client beware! Before agreeing to unusual features, establish the principles for long-term maintenance, cost and responsibility. Nurseries are rarely in a position to afford big budgets for such things, no matter how well intentioned the initial idea may be. Sustainability can be interpreted on a number of different levels.

I do not forget the private sector, which makes a significant contribution towards early-years services in the United Kingdom and the United States (much less so in Europe, where daycare and crèche facilities are generally integral to state-subsidized social and educational provision). In the United Kingdom over the past five years, major private providers have been consolidated into large nursery chains, some of which set out to provide good-quality, child-oriented environments. However, far too often the private sector, through a combination of commercial expediency and plain ignorance, will baulk at committing to a more long-term and carefully considered approach to design which will inevitably mean a slightly higher spend than they would normally get away with. When children are referred to as 'business units' is it surprising that spending on good-quality architecture is sacrificed at the expense of profitability? However, they may well argue that this sceptical approach in relation to the value of good design is fully supported by relative financial disasters like Paradise Park.

Of course, there are examples of good-quality private nurseries, usually on the scale of Burley rather than Effra; there are a number of individual nursery operators and indeed some of the larger chains have a progressive and considered approach to the environment. However, far too often in my experience, the private nursery is poor architecturally compared to its Sure Start counterpart. In a commercial world, architecture and the environment is the most obvious area for cost savings, particularly where salary costs are the greatest ongoing overhead. It is hardly surprising that costs go towards higher pay for care workers than improving architecture, a capital cost which may be impossible to recoup if the business has to fold. However, I also feel that there is something to do with the commitment of individual employees who do not have a vested interest in the business itself which militates against the investment of time and a little extra care in the spaces they inhabit, which as Burley illustrates, can make so much difference to the overall quality of the environment.

I was recently invited to come up with proposals to improve the environment within a childcare centre of one of the largest chains of private childcare providers in Britain. Although the location of the building which was due to act as a design prototype, and the grounds within which it was located, had great potential, the interior was of such a profoundly poor quality I felt I needed to walk away from the commission. To start from such a low base seemed to me to beg the question as to why exactly the proprietors were actually working here at all. One particularly troubling example was the main activity area, where young daycare children were expected to sleep while other part-time players carried on with their noisy activities oblivious to the needs of tired children. An architectural issue? Yes, as one of the key roles of an architect is to provide robust advice and guidance to the client about all aspects of the environment, which may well slide into areas of education practice and child welfare. You would not sell a car which was not fit for purpose, even if it looked good. Needless to say, the sustainability agenda was almost completely ignored. I say almost completely – amidst a sea of dedicated car-parking spaces around its entrance, one lone bicycle lock had been provided, along with a sign which read, 'You are responsible for any loss or damage'.

The sustainable nursery

Sustainability checklist (ask your architect if in doubt)

- Sustainability is a state of mind, not a last-minute thing – plan activities to emphasize the environment, with, for example, robust recycling of waste materials.
- Deal with traffic as carefully as possible to avoid unnecessary pollution – encourage users to walk, cycle or to use public transport; provide facilities for locking bicycles.
- A nursery has a duty to educate children in healthy lifestyles, with planting and gardening as part of the curriculum – design easy-access planting beds.
- Design to allow flexibility and future changes to save money, requirements can change and later refurbishments can entail high costs – plan for easy changes.
- Alternative technologies such as wind turbines and solar cells are exciting for children; they are currently expensive, but grants are available – apply early.
- Select building materials carefully – the use of timber should be from forests which are being managed sustainably and which have a declared environmental policy.
- Avoid hazardous materials in the build – paints, wood preservatives, varnishes, floor coverings, chipboards and furniture with high concentrates of formaldehyde should not be specified.
- Ecologically based landscape design can improve local air quality by absorbing carbon dioxide – maximize on planting both slow and long growing types.
- Landscape design can create important habitat for resident migratory wildlife by providing food and shelter – emphasize bio-diversity.
- Sustainable urban drainage encourages the slow seepage of storm water at peak times – specify permeable surfaces, swales and ponds (always ensuring that they are safe and secure for unsupervised children).
- Careful solar design will reduce the need for natural lighting – daylighting can be from different directions with the use of northlight and rooflights
- Electric lighting should use high-efficiency luminaries such as compact fluorescents, which have low running costs.
- Control the sun by careful consideration of the building's orientation; avoid direct south sun on large unshaded windows, for example.
- Plan the spaces to ensure noise is not a problem; although no statutory standards apply for early-years environments, acoustics standards do apply to schools, which should be adopted as a minimum standard in what can be a very noisy nursery (refer, for example, to *Building Bulletin BB93 – Acoustic Design for Schools*).
- Provide good ventilation by way of windows which are readily openable; however, always be careful of noise from busy main roads (and watch out for security). Higher spaces encourage good ventilation as long as the hot air can escape via high-level windows or ventilation shafts – ask your architect about the 'stack effect'.

A child's drawing of the exposed pipe-work in the bathroom of this kindergarten in Switzerland.

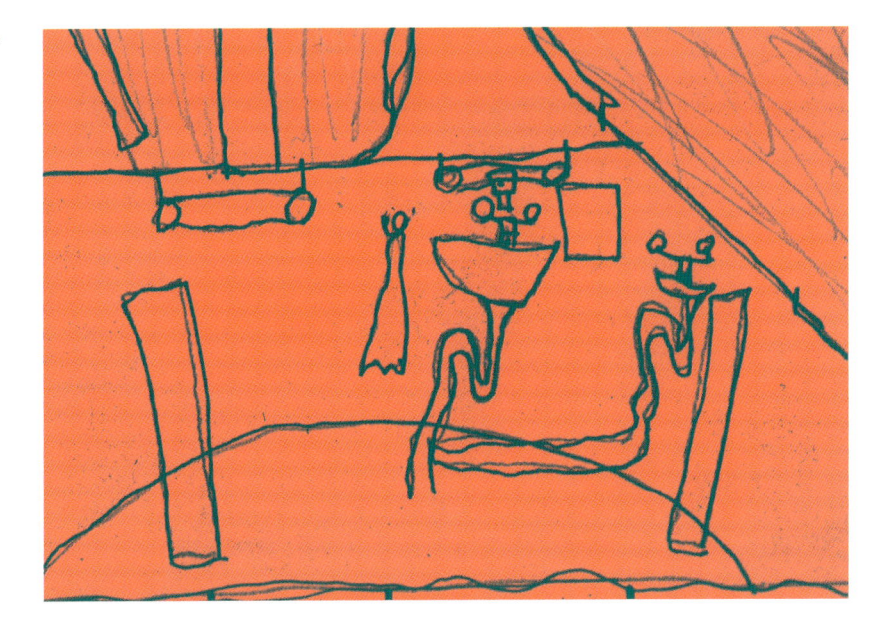

- Ensure the heating system is effective; underfloor heating avoids radiators and provides a comfortable, flexible environment – avoid air conditioning at all costs.
- Avoid water wastage; if possible provide a rainwater recycling system, infrared operated spray taps in bathrooms and dual-flush WCs – ask your architect about the 'grey water' concept.
- Thermostatic mixer valves should be used to locally blend the hot water in children's bathrooms and other child-accessible supplies so that it is delivered at a safe maximum temperature of 43°C.
- Limit the flagrant use of electric lighting; photocell sensors can be used to control perimeter lighting with movement sensors which only turn on when there is someone in the room.
- Consider the educational aspects of design. For example, a section of wall finish can be left transparent to illustrate the structure behind – it might be interesting to expose services where appropriate to show how pipes and trunking distribute water and electricity around the building.

Most critical:

- The building layout should be designed to optimize the benefits of daylighting and natural ventilation. Fulcrum state that natural ventilation is feasible in rooms with a width of less than 7 m; if any deeper, provide openings in both sides to encourage cross-flow ventilation.
- Design for control by the users so they have the power to open windows, override heating systems so they can be turned up or down if required. Fulcrum state that energy meters on display in communal areas which demonstrate how much energy is being used by the nursery are valuable for young children and staff, perhaps with an explanation of what can be done to conserve energy. Energy management systems should provide

the users with knowledge and understanding of how the building functions, ameliorating between heating and cooling.
- Take an integrated approach to create a truly sustainable building that considers the whole lifecycle of the building, which means, for example, that it is not just the initial build cost which should be considered; if expensive solar panels save enough money over ten years in electricity bills, then the initially high capital costs will be amply paid off over 20 years. Fulcrum state that serious efforts should be made to cover the environmental, financial and social dimensions of decision making (since this section was written Fulcrum Consulting have become part of the Mott MacDonald Group, specialists in sustainable design).

Summary

It is widely recognized that the critical ages 0–4 years are periods when children's physical and mental faculties develop at a startling rate. From the earliest years, all children explore their psyche through play. It is an old adage, but also an apposite one to make the observation that if a child is not allowed to be a child, he or she will remain a child later on in life. Conversely, if childhood is played out in a happy spirit of exploration, with rich and practical activities at every turn, the child will be a happy and fulfilled adult. Therefore, when children spend long periods in a daycare setting, the capacity for that environment to promote and stimulate play and exploration in a safe way is critical. Even short periods of time in play groups or sessional early-years education settings should provide a stimulating and enjoyable experience – a positive taster for the world to come, perhaps. And in this regard, the building and its related external play areas, its spaces and opportunities for play are what makes a nursery.

The tensions implicit in this statement, the relative need to be free of adult supervision so that they can explore imaginative possibilities, freely, both inside and outside, as against the obvious requirement to keep young children fully supervised and safe from harm, sometimes seems like an intractable quandary for those who are responsible for design of that environment. Undoubtedly this is where the environment can be a key driver for change, or alternatively a negative influence, actually preventing children developing to their fullest potential in some situations, and making children feel unloved – 'Why does my mummy send me to this place?'. These tensions lie at the heart of early-years design, and it is a critical issue to address. American psychoanalyst Bruno Bettelheim explained it well when he made the following aspirational statement:

> The child's environment has to be organized in such a way that it not only transmits to him, on both a conscious and an unconscious level, the assurance of being secure there and now, but it also has to transmit to him the sensation that venturing into the outside world does not constitute risk, while the future, though difficult, holds success in store for him not failure.[12]

Inevitably it demands ingenuity and a great deal of patient negotiation with the end user by any designer of a childcare environment, particularly where an existing nursery building is being replaced by a new building. Where possible, consultation with the end user (including children themselves) will help in this knowledge build. The designer of early-years environments needs to be a very particular type of architect in order to deal with the nuances and sensitivities of these users. Preconceptions need to be tested ahead of the build; one of the hardest activities is to communicate the type of building they will end up with to users. It is hardly surprising that a project such as Burley, which has evolved in a very modest way, step by careful step, with an interior which has largely been crafted by its staff with carefully choreographed displays, decorations and signage, is so successful. They have a degree of control and are empowered by the close relationship they have to its existing spaces. As a consequence there is something very personal and homely about its ambience. It is this homeliness and in a sense its anti-institutional feel which makes it so successful.

New buildings have the same duty to empower people rather than make them feel institutionalized and enclosed by rigid structures, and one of the keys to this is consultation. Making the building playful and child-oriented enhances knowledge, understanding and empathy about the environment within children who use it. This function is a job for all of us who value children's rights. As John F. Kennedy stated, 'Children are the world's most valuable resource and our best hope for the future'[13] – they need to be treated with care.

Sustainability is an essential ingredient in trying to define and explain what a nursery school is and how it should be constituted. Global awareness of environmental issues now goes beyond the basic idea of saving energy because of its dwindling supply, important though this will become over the coming decades. Designing in a sustainable manner includes perhaps less tangible and more philosophical issues such as promoting health and alleviating stress through the careful consideration of materials which are non-toxic. We must remember that a good environment is a natural environment which can be every bit as restorative as a visit to the doctor. Often it comes down to an instinctive feel for a space, it just feels right and invariably that will relate to a rigorous and careful adherence to the sustainability checklist and other holistic issues relating to a respect for planet Earth. Without wishing to sound like a reconstituted hippy from the 1960s, this is about recognizing the cosmic balance between man's development on Earth and nature itself and a thoughtful recognition of the need to bring humanity back into balance with nature. Where else would this be so fundamentally important except than in a building for young children, where the future is really all that matters?

Finally, it is important to make the point about the nursery and its distinctive qualities which means that it needs to be self-contained, with its own gardens and internal environment. At a time when many developers pay lip service to environmental distinctiveness, often converting former school buildings or, worse, temporary 'Portakabin'-type buildings into nurseries, it can't be stated strongly enough that for a nursery to work it must be designed. By that it means that all the aspects noted on the checklist below need to be addressed, so that the primary function of the children's environment works on the child's level, and not on some other adult-oriented agenda.

The sustainable nursery

The sustainable nursery school is much more than a building. Primarily, it is about the people who are responsible for using it. The building or the environment is what follows from their ethical needs and progressive aspirations. The knowledge, enthusiasm and courage of a responsible community of users determined to create an environment which fits their needs is the key defining characteristic of the sustainable nursery. The engagement of people in the process of designing and procuring new nursery school environments and, most importantly, how the environment develops further during subsequent years of use, is what we are about. This can range in scope from a cultivation patch developed in the school grounds to a multi-million-pound sustainable school designed to the highest technical standards. However, the first five years of any building's life, particularly a school, is the minimum time frame required for the users to adapt and grow their environment so that it fulfils its sustainability principles to their full potential.

The Lusterian Kindergarten designed by Konral Frey: from the rear it appears to be cut into the landscape (left), and from the front an integrated slide adds a playful aspect to the architecture (right).

CHECKLIST 2: KEY ASPECTS OF BUILDING FOR YOUNG CHILDREN

1. Do the parents and children understand the architecture? Effective consultation with the users should happen during the design and development of the new facility.

2. Does the architect understand the curriculum? For example, does he ask questions like 'How do you store things?'. This is important for effective child development.

3. You do not necessarily need lots of money to create a great piece of children's architecture. Good management and organization within the community can create beautiful environments on a shoe-string budget.

4. Don't always go for a total re-build. Often, working with existing buildings will ensure that community remains stable and intact. Too much change can be a bad thing.

5. An understanding of what sustainability means to you will aid the development of a green building which will hold important lessons for subsequent generations of children.

Chapter 3

The natural child

Under a crab apple tree ...

Playing within an adult world

> Since last September, when a group of professionals, academics and writers wrote to the *Daily Telegraph* expressing concern about the marked deterioration in children's mental health, research evidence supporting this case has continued to mount. Compelling examples have included Unicef's alarming finding that Britain's children are among the unhappiest in the developed world and a report from NCH, the children's charity, of an explosion in clinically diagnosable mental health problems among children.
>
> We believe that a key factor in this disturbing trend is the marked decline over the past 15 years in children's play.
>
> Playing, particularly unstructured, loosely supervised play outdoors, is vital. It develops a child's physical coordination, facilitates social development (making and keeping friends, dealing with problems, working collaboratively) and cultivates creativity, imagination and emotional resilience. This includes the growth of self-reliance, independence and personal strategies for dealing with challenging or traumatic experiences.[1]

The above letter was published in 2008, and in some senses it builds on other research findings and opinions by the great and the good, and simple common observational sense by ordinary parents who recognize the way in which things have changed for the worse. We can see all around us that there has been a profound deterioration in children's opportunities for play in all aspects of their lives. While the theme of this book is nurseries intended for 0–6-year-olds, arguably the story of children's play degradation runs from 0 to 16 (or whenever we consider that childhood dissolves into adulthood). However, I continue to make the point that if children are not allowed to be children, especially in their early years, they will never learn the skills needed to play and develop into healthy adults. That is why today more than at any other time, nurseries are so important.

I wish here to be absolutely clear that I do not view what some grandly describe as the 'digital playground', but what I call computer games,

The natural child

as constituting play in any real sense. For me its sensory limitations and straight-line narrative structure in no way compensate for the freedom and the multisensory delights that childhood play usually held in former times. For me this is a contemporary story of less and less time for parents to support their children and to share value-neutral free time with them. In addition, there are undoubtedly more perceived threats within the modern world, which puts immense constraints on their freedom and understandably encourages stressed parents to use TV and computer games as surrogate baby sitters.

However, there is another pressure on children. It is what you might describe as the contraction of childhood, where adult interests come into their orbit earlier and earlier. This is due to a diminished sense of propriety of many parents who are too open and careless about their own sexuality, what they talk about and when. This is linked to a breakdown of traditional media-viewing patterns, usually for commercial reasons. For example, enter any newsagent in the United Kingdom and provocative, highly sexualized images are freely on view on newspaper stands at child height. Similarly, many television programmes lack any sense of what might be described as traditional decency. Violent computer games are accessed by very young children despite so-called age limits. Liberalism and relaxed media laws have, for 15–20 years, exposed children to adult sensibilities. We must distinguish between what is right for children and what is appropriate for adults. Right now government are even raising the possibility that sexuality is discussed with incredibly young early-years children within the United Kingdom. Understandably so, since the United Kingdom has the highest rate of teenage pregnancy of any developed country in the world. So that innocence and the freedom to imagine is put firmly into the shadows.

Clearly, almost any feature which relates to the child's heightened sensibilities, such as touch, sight or even smell will undoubtedly provide improved learning opportunities. Eminent child psychologist Aric Sigman

Matthew and Amy in front of the games console. It is clear who is getting the most in development terms out of this.

The natural child

believes that between the ages of zero and three years, particularly when children are acquiring language skills, their brains are going through rapid development and are being physically shaped, like a piece of clay, in response to what they are exposed to. With reference to the negative effects of television on child development, he states that:

> We are told children need electronic entertainment or they get bored. It is not true. Children have an infinite ability to entertain themselves which television seems to erode. What children are exposed to under the age of seven, and particularly under the age of three, is of paramount importance. It's really the under threes we're most concerned about.[2]

Whereas television executives would perhaps argue that television helps children to get interested in the outside world, Sigman's thesis is that it is positively harmful, fast-moving images inhibiting their ability to sustain attention, a lack of direct adult interaction closing down social development. It is the outside world itself which gets children interested in the outside world.

Amy plays in the park with her buggy. By doing so she is experiencing the natural environment at firsthand. This top-down view of a childcare centre in London shows children separated from the external world, playing within a contrived and inflexible internal environment, which feels prison-like.

81

'Play: it exerts no external pressure to conform to rules, pressures, goals, tasks or definite action'[3]

The digital environment is not play. So what is play, and how can the designed environment promote it?

For me, play has to be about the physical exploration of the environment within which they live, and if children spend much of their time in nursery that environment has to be open to exploration in diverse physical ways. Arguably, this is what has been lost in the space of 20–30 years from the lives of many, if not most urban children. The psychological openness to the world that most (but not all) children have in their early years, gives them great potential. If they are allowed, this perspective on life makes children excellent at fantastic, imaginative games. It seems that many children are more in tune with the spatial subtleties of the landscapes they interact with, and that unique freedom of imagination which most children are blessed with gives them the opportunity to make virtually anything of their environment. However, it needs to be a particular type of environment for the best type of play.

In 1976 child psychologist Dennis Wood carried out over 500 hours of observational study of children in what he called the backyards of his home town, Raleigh, North Carolina. His aim was to discover how children played in these largely undesigned settings. He recorded what he had learned by watching kids play in the dirt, splash in the water and mess in the mud. These were children playing in largely undesigned spaces with very few purpose-designed tools of play such as toy dolls, teddy bears and train sets.

What is revealing about these observations is the extent to which children fantasized and dreamed within a very loosely supervised setting such as the scrubby backyard. Over his 500 hours, Wood identified an amazing 50–60 different thematic structures played out over a sustained period of play, invented totally by the children themselves. There is a particular aspect to this which I sometimes think we have lost from our nursery spaces in recent years; it is the nature of a type of space which the backyard comprises which is organic, earthy and loose fit, with largely undesigned malleable surfaces such as puddles and shallow pools of muddy water. By comparison, most modern play areas are rather fixed and rigid, maybe covered in the ubiquitous 'wet pour' safety surfacing. Instead, Wood's subjects chose muddy overgrown edges of these largely organic areas in which to act out their fantasies (although it is not at all clear that they have any alternative). The favourite place the children generally chose to use was beneath a crab apple tree surrounded by a shallow layer of sandy loam, itself bounded on one side by a lawn and on another by a field of dark humus-rich clay.

Indeed, it reminds me of my own experiences in my grandmother's garden, which comprised a number of manicured lawns which were separated by planting beds with narrow stone steps (to get to the roses without treading on the soil). As a child I loved walking across this terrain, which could be imagined as a chasm between two ledges or a swirling river, or anything else the child in me conceived it to be. The point is that it was natural and capable of being manipulated rather than being fixed and man-made. But listen to the charming if meandering thoughts of children M and G, recorded by Dennis Wood:

M and G had just arisen and were outside waiting for breakfast. G age two, was fully clothed but M, age five, wore only a pair of jockey shorts. M said, 'We're going to play STEAK!' M picked up an old refrigerator shelf and placed it on the grass and surrounded it with old bricks, clearly simulating a grill. M said 'Charcoal' several times. Picking up some pieces of broken slate from an old roof M said 'These are the steaks.' He puts them on the grill. G takes one off, but M puts it back on. 'Now we need some chocolate milk,' said M. Finding an old can M used it to carry water out from the kitchen. While he was gone, G moved the steaks around on the grill. Having returned, M scraped a shallow hole in the dirt with a piece of slate. He said several times, 'Now we're going to have some chocolate milk.' M poured the water into the hole and then put the dirt from the hole into the water. G watches. The water disappears, and M tries again to make chocolate milk. And again, G seems fascinated. After the third attempt I shouted over. 'Where'd the water go?' M replied, 'into the dirt.' M turned to the grill and moved the slates around. He said, 'it's the glue for the dirt,' and then almost instantly added, 'It's a swimming pool and this is the stream,' at which point he dribbled the rest of his water into the hole. He seemed to be talking to neither G nor me, but rather describing the on-goings to himself. G starts mounding up the dirt near the hole. M shouts 'Its meatballs and spaghetti,' and suddenly starts wadding up the damp dirt. Then he says, 'Its mudballs and peppers.'[4]

A page from Dennis Wood's research paper.

Wood believes that the important dimension in this play is the fluidity of the materials available to the children: sand, clay or dirt were materials which could all be moulded, whether dry or wet. This gave the children scope for imagination because there was no pre-conceived adult image attached to the children's play tools. Broken roof slates are steaks and muddy water is chocolate milk. It is an abstract canvas within which anything can be imagined, exploiting a state of mind when young people are at their most imaginative. A similar fluidity could be discerned in leaf piles, snow and water in more recent research carried out by Alison Clarke at the Education Institute in London. Providing the children aged 3–5 years with disposable cameras, Alison asked them to go around their nursery environment, inside and outside, taking pictures of their favourite places. For example, an image of a tree trunk on the edge of the play area offers a particularly interesting place for at least three of the young researchers. It may be that the fluidity of the leaves and the slightly warm, unusual and, of course, naturalistic touch of the tree bark was the real key to their endorsement of these particular areas of the nursery. However, it does also suggest to me that many children have an intuitive relationship with the natural environment if they are allowed to spend time there.[5]

Architect Christopher Day has a similar view and believes that the hectic lives of many children (and the resultant stresses) can be addressed by play within naturalistic settings. Referring to his Nant-y-Cwm Steiner Nursery in Llanycefn, Wales, Day talks about how many travel long distances to the centre by car. This deadens their range of sensory stimulation as they experience a kaleidoscopic fast-moving largely visual view of the world which lacks broader sensory stimulation. Therefore his nursery environment provides a therapeutic journey across the landscape which helps to put them back in contact with a more natural, slow-moving, fully sensory experience of the world:

> They have therefore about a hundred yards of woodland walk, crossing several thresholds to leave that world behind them. First a leaf archway, then a sun-dappled cliff edge above the shining singing river, then shady woodland, then pivoting past the firewood shed, through a gate to a sunlit play-yard and sand heap. Then an invitingly gestured, but slightly asymmetrical, so not too forceful, entrance. Then a blue-purple-green corridor, quiet, low twisting, darker. Then again a sunlit stopping place for de-hooting.[6]

Amy plays in the dirt beneath an oak tree, despite the concerns of her mother regarding hygiene as she checks the smell and texture of tree bark by sucking it.

The natural child

While full of a somewhat nostalgic longing for the more naturalistic childhoods of the past, this analysis is an incredibly sensitive reading of the environment, with some man-made features, some natural, but both parts complementing each other and creating an enduringly strong memory of place for children to carry through into the rest of their lives.

The entrance threshold at the Nant-y-Cwm Steiner Nursery in Llanycefn. Refer to Day's excellent book, *Environment and Children: Passive Lessons from the Everyday Environment*, Architectural Press, 2007.

The turf-covered roof and discreet 'garden' wall which follows the line of the road invest it with a secret, almost mystical quality.

Inside there are no straight lines, everything is cave-like and organic, reflecting Rudolph Steiner's educational philosophy.

The building's form and textures replicate the natural forms of the woodland setting, a gesture appreciated by children.

SECTION AA

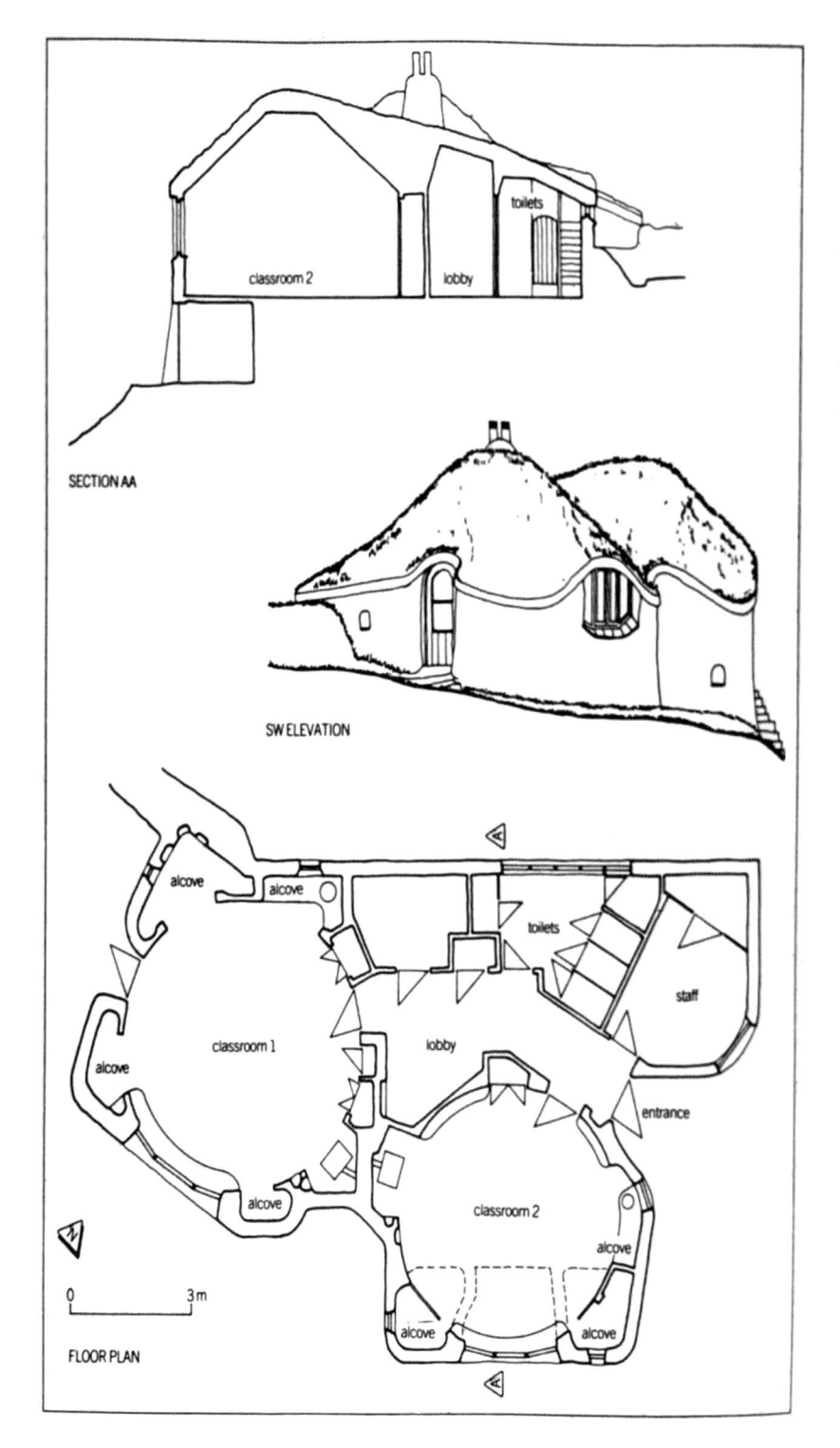

SW ELEVATION

FLOOR PLAN

The building is very small, containing two activity spaces for 12 children in each. The designers have used a circular form for each, which is justified on the basis that it is a more social or democratic form than a square space; the circle also reflects conventional early-years curriculum modes such as 'circle time'. However it is not a pure circle as that would be against nature.

Childcare expert Harry Heft suggests that environmental features should be described in terms of the developmental activities they encourage.[7] He calls this concept 'affordance', in which, for example, a smooth, flat surface affords or encourages walking or running, while a soft spongy surface affords lying down and relaxing. A room full of light and shadows has the potential to stimulate children in a particular way, whereas a dark warm place with soft furniture, lots of cushions and spongy flooring can calm them down and provide a suitable setting for storytelling, resting and/or sleeping. The atmosphere of each place within the nursery has its own role to play in constructing a suitable child-centred narrative; even entrances, washrooms and circulation areas have affordances if designed, which provide children with the opportunity to explore and interact in an imaginative way, enhancing their social skills and thus promoting their understanding of the wider world. I suspect that Heft would argue that even a storeroom, as long as it is safe and accessible to young children, gives them the chance to play and learn in a space, which would normally be out of bounds. To them it is a privilege and imparts responsibility.

Of course, one of the problems for many modern children is the lack of outdoor spaces within which they are able to freely interact. We have already mentioned the negative effects of television and other digital entertainment technologies. However, the culture we are in also helps to limit play opportunities for young children. Fear of accidents and stranger-danger predominates, so parents and carers become over protective, restricting the child's freedom to explore their environment unsupervised (or at least with the allusion of aloneness) as they used to in former times.

On a personal level, my childhood free-time living in a medium-sized town in the north of England comprised incredible freedom to explore alone or independently with friends, really from a very early age. Initially as a four-year-old, the street and rear gardens of friends would define my public life beyond home. Later on, from age six or seven, the local fields would take me 1–2 miles distance from home. Aged nine, I would take off on my bicycle and travel up to eight miles away. My epic bicycle rides out of town to visit various stately homes in the country comprised my disappearing for entire days during the summer holidays. I'm sure that my parents were completely relaxed about this independence, although if they had known exactly what we were up to, they would probably have been a little more circumspect. Today, there is no possibility that I would afford the same independence to my own children.

The manager of an early-years facility recently explained to a colleague of mine that 15 years ago her daughter had fallen in a schoolyard and damaged her front teeth, and she never wanted another child to go through the same experience. Because of a freak accident a long time ago, a generation of children were being forbidden any physical activity or challenges within her nursery. Similar anecdotes too numerous to mention paint a picture of a world fashioned in overly proscriptive adult terms (with the safety agenda to the fore); effectively this closes off to the child potential for rich character-building experiences where they interact with an unknown world firsthand. Take, for example, that most inviting world for sensory experiences, the kitchen. This is a room which arguably is part of a sequence of ordinary

The natural child

domestic spaces in the home to which most children have access without question. In most institutional settings, it is out of bounds. Play kitchens abound, but access to real ones is not permitted, which creates a false boundary between servant and served; some argue that limited safe access for children can be facilitated by careful design and good supervision.

A kitchen integrated into the activity area. Ideal for nurseries with fewer than 20 children, reproduced by permission of CNAF – Fondation de France Ministere Des affairs Sociales, published 1994, by NAVIR, 'Les temps de l'enfance et leurs espaces'.

A play kitchen in a London children's centre.

The natural child

An activity kitchen in a Danish kindergarten. Note the two different floor heights, one side for adults, one side for children.

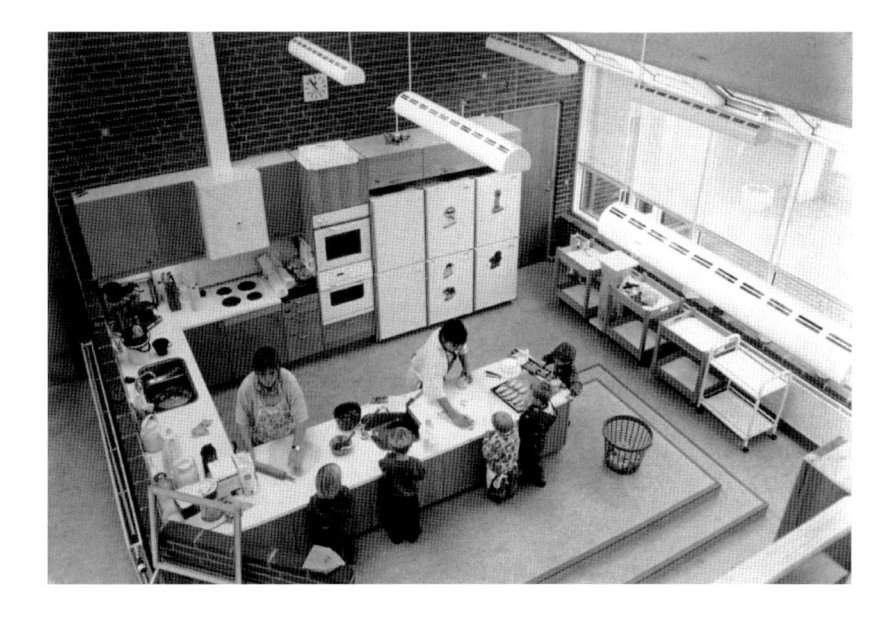

Whereas European childcare systems were largely based on education, the UK childcare system has its historical roots in the child welfare movement, catering for vulnerable or at-risk children. Even today this culture predominates. The training of childcare workers, and the health and safety legislation that governs daycare nurseries, emphasizes children's vulnerability while tending to underestimate their ability and competence to interact with a complicated and sometimes challenging world – in particular, their capacity to negotiate the physical world where the effective development of body and brain coordination is a direct function of taking risks (and sometimes getting hurt). Undoubtedly this conundrum is a challenge to which the nursery designer must rise. How can you introduce children to a world without risk? Furthermore, we are in danger of creating a surveillance society where children lose the idea that over and above everything, they are individuals within a free society and can if they so choose, disappear off the radar and operate freely and in private. More than ever, the modern children's centre needs to replicate the challenges and the broad range of sensory stimulations which in former times used to be more freely available to them. At its best, the contemporary children's centre must be a microcosm of the world.

Mary D. Sheridan, a senior community paediatrician who wrote a series of very influential books on children's play during the 1960s and 1970s, defined real play as the opportunity for children to play spontaneously in a framework which was not determined by adults.[8] What is distinctive about this is the type and duration of play. Rather than being limited by neat periods of time determined arbitrarily by adult time frames, it should go on for as long as the child wants it to. Even if the session ends and parents turn up to collect their children, the nursery should have enough space for the game to remain in place until the child returns for the next session, when it may continue. If it is entirely determined by the children themselves albeit constrained or promoted by the environment in its totality, its value is perceived to be far higher than play which is overly prescribed by adults and governed by strict time frames. Children can choose when to change their games and when to stop and switch to another, different kind of play. However, in Sheridan's terms, it should always be down to the children themselves. In this context there is no obvious outcome to the play activities. To the child, playing is an end in itself, the only reason for being in nursery. Given the prescriptive nature of most nursery activities today, it seems like a somewhat fantastical view.

Sheridan identified four distinct aspects of development that play might promote in children. She defined these as 'apprenticeship', 'research', 'occupational therapy' and 'recreation'. She observed that as children get older they develop competencies in performing everyday tasks which are honed by play. So, for example, at age 2.3 years Amy simply allowed herself to be changed and dressed by her adult carers; however, she was watching closely and understanding the nature of the activity during this time, whereas previously she would have been largely oblivious to what was going on. When Amy was 2.4 years she began to dress her baby doll in one of her own nappies, and although she found her inability to do it herself immensely frustrating, with mummy's help she could dress a small doll in a huge nappy and put it to bed in her own little bed. This particular activity became an important dimension of her play for 2–3 weeks during this time in and around the home. When aged 2.5 years Amy found the sticky tabs on the side of the nappy she was wearing and began to remove her own nappy when she went to the toilet before bedtime. Finally, at the age of 2.7 years, she was not only capable of changing her own nappy, she could dress her doll and began to dress herself, albeit still exhibiting profound levels of frustration. It was for me a simple yet gratifying sign that her play patterns were healthy since she was keen to try new things out well before she was capable of actually doing them for herself. Her skill and competency was advancing as she played. The challenge of play was never too daunting to put her off because the environment was so nurturing. A parent should of course always enable this to develop rather than restricting it. This is what Sheridan describes as 'apprenticeship'.

The second category in Sheridan's list was 'research', which she describes as a 'process of observing, exploring, speculating and making discoveries'.[9] For example, for the child learning about the properties of water by testing different objects in the bath, something that floats and something that sinks will be very different to touch and handle. For Amy at the age of 14–16 months, her doll's house with its doors to open and close and door openings to push characters through, Noddy's car or the Little Pony acted as a primer for testing the real thing at the age of 1.6 years. Although restrictions were imposed on playing with real doors in and around the house, Amy found a garden gate to play with on the way to the shops, which – with daddy's help – she could push open and pull shut. While daddy is shouting 'fingers!', by the age of 2.2 years the gate becomes a very real play activity. The gate separates one kind of space from another type of space, the street from the garden, but it is reasonably safe, and the owner of the garden is relaxed about it all. Amy goes in and out for at least ten minutes. It is a game which is tantalizing and exciting at the same time. According to Sheridan this activity would come about as a result of her own research. So engaging is it to Amy that the only way for daddy to move her on is to carry her off

Illustrations of Sheridan's four development criteria: apprenticeship: Grace learns basic skills by watching her older sister (top left), research; Grace is learning about the properties of water (top right); occupational therapy: Amy is upset, she is calmed by the introduction of a special toy, her concentrated play distracts her from anxiety (bottom left); recreation: children simply playing and having fun largely unplanned and un-directed (bottom right).

physically, while she screams. It is an illustration of how a certain type of play with toys can then be translated into experimenting with the same concept in a real setting, with full-size doors and thresholds separating different spaces. Play through research provides ample opportunities for this kind of informal learning.

The third category is called 'occupational therapy' and refers to the type of play which is intended to console the child during times of boredom or emotional distress, or even physical pain (although this seems a little unlikely since most young children will wail their heads off in the face of any sort of pain until they are physically consoled by an adult carer). Here, the thought I have about this in relation to Amy's current behaviour is that she does have certain therapeutic strategies which often entail a change of room or some sort of movement within the space. For example, at bedtimes she will ask to go to the toilet and to brush her teeth, activities which change her environment and give her an alternative perspective at a time of some stress when she is about to go to sleep and determined to avoid the long period of sensory deprivation. Similarly, she will ask to be chased around the dining room table at certain times when a new adult enters the room in the middle of a long indoor session of activity. This game entails running after her and then reversing the process and being chased by her; it is clearly important to maintain clear pathways around the space so that this can occur without accident. It is a refreshing change for her and in a number of similar instances a more physical form of play which can lift her mood.

The final category which Sheridan describes as 'recreation' is the function that most readily springs to mind when we think of play. Children entertain themselves through play for its own sake – they are simply enjoying themselves and having fun. Here, active play with other children often involving falling over, and tumbling springs to mind as a particular dimension of her development. On a basic level it is a means of gaining strength, agility and confidence in her coordination and balance, developing brain and body coordination to give her the skill to negotiate the physical challenges of the nursery. It is interesting to observe Amy now at the age of 4.4 years, with her younger sister Grace (2.6 years), as they play together. Amy is leading the play as they hold their Barbie dolls, each in close proximity. Their voices are audibly high, as if to project them onto the dolls like ventriloquists; the children are acting as the parents, supervising the hair wash or the school run. In Sheridan's terms Amy is engaged in 'occupational therapy', Grace is the apprentice.

In reality each role is so overlaid with different aspects of the four criteria that Sheridan's narrative feels ever so slightly simplistic. How far it is possible or even helpful to rationalize these types of play down to four items is open to question. One might argue that modern life is nowadays so complicated that all sorts of play overlap and intersect in any event to reflect that cultural dynamism. What we can agree with is that it is essential for the child's development and there is no single type of play:

> Play is as important for a child's developmental needs as good nutrition, warmth and protection. It provides opportunities to improve gross and fine motor skills and maintain physical health. It helps to develop imagination and creativity, provides a context in which to practise social skills, acts as an outlet for emotional expression and provides opportunities to understand value systems. Providing for play includes ensuring that the child has opportunities, resources and time for play appropriate to her stage of development.[10]

More contemporary definitions of play can also be found in the book *Play in Early Childhood*. These set out to break things down into outcome-based analysis as follows: *physical play* is defined as that which increases the infant's control over her movement. What is interesting about this is how gender differentials emerge from a relatively early stage. Often, boys are more physically active than girls, more adventurous in climbing and other risk-based physical activities. Girls are generally better at inhabiting space in a way which combines movement with social interaction; for example, two activity corners and the bathroom will form part of a sequence of movement routes around the nursery for three girls in play together and perhaps an area outside in the garden will form a fourth area. Often, boys love the gym, where they can rush around and fall over, their engagement with their own physicality being much more instinctive than girls. While it is only anecdotal, my observation of three-year-old Amy and her three friends in nursery appears to exclude boys from their play.

The distinction between these two play types is termed the development of fine motor skills and the development of gross motor skills, the former being activities such as tying shoe laces and painting, the latter relating to activities such as jumping and throwing. If the nursery curriculum promotes one over and above another, children will develop in uneven ways. Spatially the implication here is that designers must be careful to provide spaces which are not disrupted by the physical activities of other children. Free access out into the garden is important, with safe run-in–run-out play and level thresholds. Quiet spaces as well as loud spaces for boisterous activities are equally important. What may be deemed as purely functional spaces such as bathrooms and storerooms can also be exciting places for exploration and development if given the right architectural tweaks. Everywhere is potentially a place for learning.

A second definition which is described as *cognitive and symbolic* for me sounds really interesting as it hints at a more off-beat type of understanding, which I earnestly believe most children are open to; just as they are open to the fantasy of a Harry Potter story, for example, all founded on flights of mystical fantasy and extremes of the imagination. However, the description in the book is a little more prosaic, being the accumulation of knowledge, 'cognition' and the development of skills for using information, problem solving and determined action based around 'symbols' around the nursery environment. For Amy this stage is currently illustrated in her instinctive wish to do everything herself within reasonable possibilities. So, for example, I have learned that she must be able to switch her overnight light on at bedtime herself – if I do it she tells me to switch it off (in her own special moany way), so that she can do it herself. These actions and activities spring from her conception of the space which she inhabits. The more control she has over the space the more empowered she feels. This asserts her individuality and her sense of herself in relation to the environment. Therefore, you can provide a nursery environment that is full of buttons and switches at child height (which may not necessarily do anything), yet gives the child a similar sense of control over their own environment. The more control you give to the child over their space, the better. I firmly believe this is an important aspect of early years; however, I am sure that many childcare professionals would not go along with such an anarchic view.

The term 'symbolic' is used in conjunction with the child's development of *linguistic skills*, learning to talk and related forms of expression. However, most important is the sound of happy adult voices; this gives children the sense of language, and most importantly the joy of communication. Here, if the correct ingredients within the child's world are present, language emerges in surprising waves of comprehension mixed in with confusion and inevitably some complete gibberish. However, the environment has a key role to play in this if it is correctly choreographed.

To reiterate, at the time of writing Amy is barely 2.4 years old, yet the words and phrases she uses are developing at pace, often tumbling out with such amazing dexterity and spontaneity it can at times sound like a form of street poetry. At 2.2 years Amy was doing a lot of what I view as pretend

The natural child

talking, where the tone and melody of adult discussions are simply mimicked. Then nouns begin to emerge relating to her own actions, such as potty, dummy, nappy, water, porridge, rain, door, cat, baby (she has a little sister), shoes, coat, hat, buggy. Observations of street life are full of distinctive objects which for Amy were the foundations of her language development when she was 18 months or so, such as car, bus, post office, shop, dog, bicycle and coffee. The everyday urban environment can be far richer for language development at this stage than special places such as the zoo. Even later, at age four, she barely notices the lions as they drowsily wander around their compound; she is much more interested in the gift shop, with its shelves loaded with toys, books and sweeties. It almost acts as a memory bank, the novelty of new images for play laid out and ever-changing as she moves around the shop space, just as she would move along the high street in Notting Hill.

At this early stage there are few verbs, just possessive adverbs – especially 'my'. As a parent carer tending to be around some mornings and most evenings, I find the home environment with its suite of dedicated rooms for particular functions is a particularly rich source for language development. Lets not forget that most language is relating abstract sounds to images, which if repeated a few times will lodge firmly within the memory. So very simple, a window with a view to the garden has all the following words and many others: bird table, glass, grass, shed, sky, stars, rain, night-time, swing, that is beyond the room; then there is the foreground, where glass, floor, ceiling, wall, picture, chair, lamp, skirting, wall light, shelf (and special items on the shelf), form a memory structure which is contained by the function-oriented space. Anything you can point at and say 'what is that?' is a lesson for later which is repeated everyday until the child's vocabulary is full. In this respect, I believe that it is positive within the nursery to have as much as possible on view; however, it must have an ordered complexity which does not overwhelm the child's curiosity in a chaotic way. Another point is that rooms within the nursery should each be distinctive and rotated, rather than being used for a single group of children only. Nursery designers can learn greatly by observing the ordered variety of the domestic home.

The fourth category cited here is *emotional and social development*, and is the ability to relate to a group. It is a set of skills and understandings which are initiated in the home, but they can then flourish in nursery, particularly important being the facility for social interaction between children, what the Danish call 'social pedagogy'. Sometimes children do not receive consistent kinds of discipline within the home and will arrive at nursery with very few social skills. They can be defensive and aggressive. Very quickly they will learn to be disciplined and will gain control of their emotions, finding acceptable ways in which to demonstrate opinions and feelings. Most importantly they will learn to respect other children's boundaries, not intervening in play to which they are not welcome. This is one of the key roles the nursery provides and an important learning skill for some disadvantaged children. The nursery enables them to fit in to the conventions of society. One aspect of social development which has been mentioned previously is dining communally as a group, in mixed-age groupings, with older children helping to serve out food if possible, and of course assisting with the tidy-up at the end. It is such an important group social event full of development opportunities and it is disappointing how few daycare settings actually provide a dedicated dining area where the whole community can sit down together.

The final category in this particular list is the notion of moral and spiritual development and is broadly concerned with values that bind society together, such as honesty, fairness and respect. In today's relativistic world, it sometimes seems problematic to assume there is anything as straightforward as a right and a wrong. However, with people aged 2–4 years, right and wrong are clear values around which much of their subsequent adult life can be structured. Inevitably this overlaps with the previous section in its social dimension – being able to get along with other people is essential in any moral and spiritual comprehension. In previous generations a church-based ethos would have been the cornerstone of much of this work; however, we live in a multi-cultural world where parents with very different backgrounds bring different values which sometimes simply do not gel collectively. However, my experience working with an inter-denominational nursery group in Northern Ireland illustrated to me how most children just get over these differences; indeed, the nursery is the foundation of tolerance and a sure way to overcome generational bigotry.

To complete this section the idea of spirituality is a concept which many early childcare pioneers believed children had in buckets, and if they received this from an affinity with nature in and around the nursery, then the process of education would be more natural and spiritual. This is an area to which we will return in the next section.

Play outside

Mary Sheridan's findings emerged as a byproduct of her work as a child paediatrician. As must be fairly apparent, my current observations come about largely as a result of being able to watch my own young children interacting with the world as they grow up, with half an eye on this publication of course. It is one of the few benefits of being a middle-aged father that I seem to have a little more time to spend with them than most younger parents. It is not exactly using my own children as a research project, but it is surprising how quickly things seem to move on for them in subtle but quite profound ways at this time in direct response to certain key environmental stimuli. If you really spend time with them on a daily basis it is knowledge which is profound.

At the time of writing, Amy is 2.4 years old and looking at what she does in a studied way, how she uses her environment as the drama-laden way in which she develops her competencies and how her language development is tied so closely to things and objects within her physical landscape (both inside and outside) is a real eye-opener. Indeed, it is her ability to put her names to different features of her environment which is particularly empowering in this respect. This could perhaps be related to the naming of inanimate objects of the everyday with poetic or otherly names. For example, a Jaguar car or a Mars Bar, whose iconic understandings have no relationship to the object's real meaning. If known in childhood carries through to the rest of most people's lives such is the power of a name. Some cultural geographers theorize that the naming by early explorers of frontier landscapes in the United States allowed them to internalize the unknown territories as places. The specific names used, such as 'mount', were often imported from their own experience and culture far away, and it would have little relation to the actuality of the terrain. According to Yi-fu Tuan, 'speech is a component of the total force that transforms nature into a human place'.[11] In the same way, a child often uses poetic language to absorb and understand their relationship to a particular place in the nursery.

This underlines to me how strategically important it is to structure the nursery environment with the right levels of sensory stimulation. It should be neither too sparse or too busy and over-decorated. It should not be too abstract and rigid as to allow little change by children themselves, and the naming of areas and places within the nursery which have a particular character, dens if you like, should be encouraged, just as each child is given a distinctive name. Names for places can empower children and help in language development, also providing the nursery environment with a richness which transforms the building into a romantic learning landscape of the mind. There are keys to this structure which we will come to later. It is also very clear to me what a catalyst to her development any change in environment can be, such as a trip to the seaside or the countryside, where all sorts of experiences accelerate her curiosity and development at an almost logarithmic rate.

You may have noticed that the context of much of what has been described so far in this section on play has been outside spaces, or at least those which have their emphasis on the naturalistic. From the kaleidoscopic wooded threshold of the Nant-y-Cwm Steiner School by Christopher Day to the backyards of Dennis Wood's home town, Raleigh in North Carolina, we have talked about children's play largely in the context of what happens outside. It is important to emphasize how critical it is to provide outside play spaces for children in any nursery setting; however, I argue here that the nature of that space, its design emphasis, remains as important as the building itself. To reiterate, the emphasis should be naturalistic since this has a proven and positive effect on children's well-being, especially in the most urban of contexts. The addition of certain types of planting rather than plastic play equipment, the greening of the environment even in small ways, is not just about aesthetics, it is a force that re-connects children to nature.

It is hardly surprising how many childcare experts from Froebel in the eighteenth century to the present time relate mental well-being and healthy development in childhood with the systems of nature and the child's ability to interact with them. For example, Edith Cobb, in *The Ecology of Imagination in Childhood*, states how critical these relations are to young children and how the natural environment can support them in a way almost nothing else can match.[12] Susan Herrington and Ken Studtmann, who carried out research into landscape oriented play yards in Iowa, noted how

The natural child

Children often use poetic language to absorb and understand their belonging to a particular place in the nursery. Here, generations have used this tree in a left-over corner of a nursery in Westminster, London, variously describing it as the 'look-out', the 'princess' palace' or the 'raven's nest'. Similarly, the circular shelter in the grounds of a Frankfurt nursery conveys a sense of place to the children to the extent that they have christened it the 'round table'.

The natural child

the placement of stepping stones in a meandering line which weaved through typically unused spaces of the playground made children use these previously unused areas. In addition they increased the children's spatial and cognitive awareness because prior to their intervention the children moved directly to conventional play structures and the spaces underneath them. The children were physically challenged by the new pathways to the point where they became actively involved in the scrubland around these new areas. The stones were placed slightly beyond the children's normal stepping pace, causing them to make a step-jump from one stone to the next. Then, working in pairs, two groups dug up and moved the large terminus stone a distance of ten feet. These actions required physical strength, cooperation and planning and exhibit a similar changeability explored by Wood in his research.

The two young researchers also observed how conventional play equipment initially helped the children to develop their fine and gross motor skills, and certain equipment, such as the mini-slide and the elevated tunnel with platforms at either end, were utilized as locations to socialize. Interestingly, boys tended to monopolize the equipment, using it to test out their physical abilities at the expense of the less physically confident children in the group. These children then seemed to use it predominantly as a place to test their physical prowess as a means to establish social hierarchy in the group. Children who were stronger, faster and able to climb higher became leaders in the social strata. Hence the social hierarchy of male children was directly linked to their physical prowess.

During the second phase of the experiment, plant enclosures were constructed of *Euonymus alata* and *Thuja occidentalis*, which formed rustic dens scaled to the height of the children. They were located in open grassy areas away from the traditional 'off-the-peg' equipment positioned on hard asphalt surfaces. The use of these vegetative rooms altered the social focus of the group almost overnight. The soft rustic dens overgrown with vegetation became the main focus for fantasy play. According to Herrington and Studtmann, these rooms were used more frequently and for longer durations than was the case with standard equipment. Dominant social hierarchies which previously had been based upon physical prowess now became based around the child's command of language, their creativity and inventiveness in imagining what the space might be. A different form of play assumed control which was directly related to these naturalistic interventions.

Plant material installed in the playgrounds added diversity of plant type, texture, colour, smell and wildlife habitats. Tactile experiences with these plants enriched the children's experience of play, especially those with learning difficulties who tended to spend more of their time in these soft, gently aromatic areas. The spatial configuration of the plants themselves appeared to have profound effects, transforming the social structure of the whole group for the better. Some planting actually became enclosures or children's dens in their own right, such as the large-leafed rhubarb plants. Specific names were even invented for the vegetative rooms; one was called 'princess' palace' and the other 'raven's nest', the names denoting specific activities and genders.

Boys tend to dominate areas of the nursery which engage them in robust physical activities. This tends to push girls to the edges of the space.

The University of Iowa experiments illustrate how a more horticultural approach to the design of children's play spaces can affect the way in which they play and relate to one another. While they spend 40–50 hours per week in institutional, commercial and other out-of-home care settings, evidence suggests that very little of that time is actually spent engaging with nature. Rather, children's experiences are of an altogether more synthetic quality, which gives them a strange perspective on life. The researchers showed that through the introduction of relatively simple landscape elements

The natural child

into these hard, equipment-based playgrounds, different types of development were encouraged immediately. The landscape-based approach was salient because it could link cognitive understandings of space with specific design layouts. The research concludes on the positive findings that the landscape external environment definitely has a positive role to play in child development in particular as an extension to the nursery building itself. However, it is clear that more research is required from the landscape community to fine-tune the types of intervention that are realistic on tight budgets and ensuring that they are maintenance friendly, so that it does not become an impossible aspiration.[13]

Susan Herrington, who was a joint author of the research in Iowa, has recently completed another research-based exploration, this time working out of the Landscape Department of the University of British Columbia. Based on five years of study, the research posited the following question: which outdoor physical factors contribute to early childhood development and help to raise the quality of play at childcare centres, and to what degree do these factors currently exist at the centres under study? Sixteen outdoor centres were chosen for the study, selected on the basis of their similar socio-economic location across the city of Vancouver. Different architectural types of childcare centre were chosen, which are described variously as modern, organic, modular and re-use (definitions which in themselves illustrates quite nicely how low in the architectural pecking order most daycare centres are). First, the average size of the space was discussed. It was found that despite increasing concerns about child obesity, the amount of play space allocated to children in both Canada and the United States has remained constant at $7\,m^2$ per child. However, because of new safety restrictions on the encroachment to play equipment, these restrictions have further decreased space for gross motor play. In other words, the area for each child is far too small and it is in any event even more restricted by large set-piece play equipment. The findings were quite clear: childcare centres that exceeded their recommended

Twenty square metres for your Land Rover Discovery, nine square metres for your child discoverer. Today, cars increasingly dominate the urban environment; a nursery which is local should encourage parents to accompany their children on foot.

The natural child in a rustic setting.

child densities experienced more aggression between children. Perhaps more alarming was the finding that expensive fixed-play equipment is not only expensive to buy, it is also vacant 87 per cent of the time.

When Local Authorities or, more often, private nursery developers have the opportunity to spend a bit of money on their play spaces, the result will often be a heavily themed play space – for example, an Alice in Wonderland theme or a Peter Pan theme using somewhat inflexible 'off-the-peg' play equipment. There is one famous play park in central London which has a full-size ship 'floating' on a sea of sand, an all-encompassing theme which dominates the entire landscape. Unfortunately, Herrington's research showed that there was no discernible relationship between overt themes created by manufacturers or designers and children's imaginative play. This suggests that the themed play area is not intended for children at all; rather, it would appear to be directed towards parents and used as a selling point, in schools and nurseries, who might well be making choices about which daycare centre their child will attend. What was much more important was the yard shape and equipment location. Most important in outdoor play spaces is that they contain materials which can be manipulated, such as sand, pea gravel, mud, plants, pathways and loose parts such as blocks and tree logs. This is a strikingly similar finding to Dennis Wood's children's play research undertaken 30 years previously. The research concluded that aggression increases when no manipulative material is provided in outdoor play spaces.

The *7Cs* research document provides a helpful series of guidance notes for those designing new outside play spaces, which we paraphrase here. *Character* defines the distinct image that each play space should itself own. It expects the design team to produce a mission statement that provides a framework to the overall image required by the group. The consultation should include professional designers, educators and parents. Usually (but not always) there is an existing space or context around which the image can be constructed. There may not be enough resources to do everything in

The natural child

the first month or year, but the benefit of a mission statement is that it can be returned to as a reminder of deep-seated character required. Character is also important to the child's development as their early memories form a cultural environment to which they will return again and again. Every sensory perception is partly governed by the cultural environment of the child. My own childhood memories are strongly informed by the trips to the country and, in particular, the river running through the Dovedale area where my parents took us. The stepping stones over the river were the aspects of the memory which gave character.

Next in the list of Cs is *context*, which of course is given although it can be modified by design. Considerations such as the relation between the play space and the surrounding area should be considered. I always think that enclosures, physical boundaries such as hedges and more usually fences are important to create the right ambience. Fences which appear too prison-like and enclosing create a synthetic, almost angry, view of the world for children. They want to be able to look out, but this transparency will very much depend on the location of the fence. The *7Cs* authors explain how a number of rooftop play spaces accommodate views of the city which captivated children, inciting discussion whether the centre is in the country or in the middle of the main business district. Orientation is also critical; as with any building design, is the space shaded and damp or open and sunny and maybe too hot? The micro-climate can be modified by the correct positioning of shading – both buildings and planting can create the correct micro-climate.

The third C is *connectivity*, an idea which I believe is extremely important since most children do not differentiate between the inside spaces and the garden; to them it is about the experience of play and the freer is the access the better. This of course works on many levels, such as views from the inside to the outside creating an atmosphere which is positive and educational. From the weather to the time of day, children gain much when the outside world is visible to them. Another dimension of this connectivity is the paths children use to move around, both inside and outside. If there is a hierarchy of paths, wide, narrow, hard and soft, different children will be able to find their level. If it is just one area of asphalt the kids with fast tricycles will dominate and push other children to the edges.

The fourth C is perhaps for me a C too far: *change*. Here, the researchers are really referring to the provision of a variation of spaces, which cater for different-sized child groups from large 6–8-person friendship groups to 2–4-person group spaces for more private intimate socializing right down to spaces for individuals who wish to get away for a while. As adults we should appreciate that being in daycare is a public experience and sometimes the need for privacy is as important to some children as it is to some adults. Another dimension of change is to do with the way plants change over the year. Experience tells us that planting in children's play areas is invariably limited by maintenance requirements. However, play spaces that incorporate a range of low-maintenance plants which change with the seasons are best for imaginative play, particularly those which have sensory qualities, are aromatic or interesting to the touch.

The fifth C is *chance*. This is an idea we have identified throughout this section as being crucial and it is obvious really. It is the possibility that children can actually adapt and amend the space in question themselves. In a sense, the ability the child has to leave their own impression on the space applies as much to the inside as it does to the outside. However, the outside has greater potential as there are messy organic activities such as digging in the earth or sand, or construction play with timber and nails can be more readily accommodated (although I have seen big sand-pits indoors). It is a quality which has been referred to as flexibility or open-endedness and is difficult for many architectural designers to comprehend as they design for permanence rather than changeability. Simon Nicholson observed how, in a museum situation, the floor was worn down most by children where they could manipulate and modify exhibits: 'Children should have the opportunity to play with space forming material in order that they may invent, construct, evaluate and modify on their own.'[14]

The researchers add to this section the notion of mystery, which they describe as the possibility to create more labyrinthine environments in which the children can gain a sense of exploration. This can be achieved with low walls and planting which slightly obscures the children's view of the landscape. Again, this can be something which the interior can be designed to promote although it contradicts the overriding urge for childcare workers to see and be seen at all times. Nevertheless, it is a very interesting concept and if there is enough space outside, is in my view extremely valuable for young children's sense of spatial awareness and feelings of independence.

The sixth C is *clarity*, which may appear to contradict some of the other ideas discussed previously. However, discussed in these terms it is a concern which relates more to small confined play areas where large play equipment occupying the centre of the space has the tendency to restrict other activities because lines of vision and routes are so restricted by its physical presence.

The seventh C is *challenge*, which refers to the physical and cognitive encounters that play spaces provide. In this health and safety obsessed culture we and our children inhabit, any physical challenge is often designed-out as a matter of course. As I have stated previously, this is my frequent experience in my design work and it is extremely frustrating. The real tension between providing stimulating and challenging play experiences and the need to iron things out and make everything so safe (it actually restricts play) is a constant challenge for designers. The researchers quite rightly point out that children will use equipment and different parts of the environment in their own imaginative ways, particularly if it is static and predictable. For example, how often will you see little play dens in children's parks with the children actually climbing on the roof rather than sitting on the inside, where it was intended? Children will test the potential of any play setting to the limit, and the key is to create gradated levels of challenge, such as log ends which are at different heights and spacings.[15]

The *7Cs* document is tremendously valuable in confirming and reminding us of the fundamental issues of play; for example, the idea that changeability is important; that views outside of the play yard (if it is the right location) gives children a meaningful sense of their position within the wider world; and that building for challenging play, while it may contain some risks for children, are nevertheless worthwhile aspirations as they are fundamental

The natural child

The labyrinth, a place for children to experience mystery, exploration and adventure within the safe confines of the nursery. The best nursery environments reflect this idea of challenge within security.

to mental and physical well-being in young minds. These are the discussions which need to be had, particularly with childcare professionals who are entrusted with safety, organization and the day-to-day maintenance of play areas. If they are to optimize play and promote challenging experiences for children, they need to have the confidence and understanding of why and how. In short, there is a case to be made that childcare workers and early-years teachers need spatial awareness training. They can be extremely and perhaps understandably conservative in their outlook (and it has to be stated, sometimes lacking in the basic knowledge of how play is viewed in the architecture of progressive design), because most childcare practice manuals determined by local and central government pay lip service to good design, emphasizing instead practical outcomes such as safety, security and hygiene which lack any understanding of the richer contextual framework.

One of my earliest experiences of visiting a daycare setting in that bastion of good practice, Denmark, was catching sight of children (albeit 5–6-year-olds) stoking a fire in one corner of the garden. I was quite shocked that this was permitted and commented that this would never happen in a UK or US setting. The response was relaxed; first, the children needed to learn about all of the earth's elements, of which fire is one; second, they have been supervised with the fire many times since they were toddlers in nappies; third, the pedagogics have very good views of the children who simply behave sensibly and logically; and finally, the bonfire in the garden is an essential aspect of the local culture and it is probably an experience that 90 per cent of the children have in their home environment every week, so it is a duty to build confidence in the children who attend daycare.

I would like to finish this section by describing another one of my projects, the Portman Early Childhood Centre children's garden in the centre of London (see pages 164–165). I appreciate that for some readers this must appear to be something of a promotional exercise, but it is not intended that way. I merely wish to use it as a case study which I know well to explain the original concept for play, and to illustrate the way in which it has changed over the first five years of its use. The three main issues which confronted me with this project were: first, the confined space available to provide a stimulating and rewarding environment for child development; second, the challenging range of spaces the client's brief demanded from their confined area; and third, the daunting health and safety agenda which the Local Authority client (as opposed to the user client) was determined to impose upon the completed facility. To reiterate, I am only discussing this development because I am close to its design ethos and subsequent testing in use.

When first introduced to the space, I was rather disappointed. On one side a heavy three-storey brick building, the nursery itself, overlooked the external play areas. On the other side was the gable end of a large five-storey social housing block. It was a space mainly covered in hard asphalt with one decrepit old swing in the centre. Staff who had direct access from the ground-floor activity areas used it to take furniture and play equipment outside on sunny days and allowed children to run around; there was very little real play happening as a direct consequence of the impoverished environment. Following a number of interesting meetings with the community of users, we hit upon the idea of breaking the existing spaces up by way of low walls, to create an outside-room concept. Each of these 'rooms' had a different play theme in an effort to fulfil the client's brief. Added to this was a new covered canopy with an upper play deck which not only provided a secure covered play area to extend activities from the ground floor nursery, but also enabled those working on the first-floor baby unit access to their own sun terrace with a stairway down into the new garden below.

The concept and basic functional ingredients were then worked up, with the external rooms adopting distinctive themes, an activity garden in the

The natural child

centre, the heart of the space, a green garden for cultivation and planting and finally a hard area for ball games and 'parachute' sessions. In order to make things interesting for the children, a number of different levels were introduced to the main activity garden. Solid disability accessible ramps were introduced which were and continue to be used extensively by children in pedal cars or pushing buggies. But the space also has many steps and level changes, a deliberate strategy to encourage movement and spatial dexterity within the infants.

There is a mini garden tunnel through the area of willow wood planting and a series of major focal points or child-oriented features, which are choreographed with some care. First, there is a stage which has a slightly raised timber deck on axis with the nursery access doors. It has a mural backdrop (by a local artist) of a city showing the River Thames, and other key London landmarks. This is a sort of celebration of the city, a reminder to children that although they may live in poor conditions themselves, they are in the heart of a world city with the fantastic potential that holds for their futures. Most importantly, there is a water feature with a cascade from the rear of the cultivation garden wall, which then drains into a mini beach. It ends up in a deep pool, all of which is tested as child safe. I saw a similar water feature in a rooftop kindergarten designed by Le Corbusier in the 1950s and always thought it to be an amazing benefit to those children who would use an urban garden like this, particularly during hot city summers. However, the continuous trickle also reminds me of my own childhood, constantly damming and re-routing streams at the bottom of our family garden, a sort of flexible play area with the ultimate fluidity of water play as its focus. The final urban intervention is a little wooden bridge which connects the upper level of an inhabited wall between this and the cultivation garden, bridging over the upper pool area.

The cultivation garden was intended to be a structured area for digging and planting, with children aided by staff and parents encouraged to

A girl stands in front of the city mural at the Portman Early Childhood Centre, Westminster, London.

engage in the growing-for-food agenda. Unfortunately the soil drainage in this area was very poor and the initial planting sessions never really took hold properly. Perhaps there was also something about a lack of will to really make the activity area work. Consequently this area was recently filled in with safety surfacing and a climbing frame introduced to extend the potential for physical play. A further amendment was made between the ball play yard and the central activity garden which comprised the removal of the wall between the two, to aid supervision between the two spaces.

The changes which have occurred since the handover of this play area – the removal of the fence between rooms 1 and 2 and the introduction of the climbing equipment in what was intended to be a cultivation garden – are absolutely in the spirit of the scheme's original design concept in that it should not be an environment which is fixed, but rather one designed to work in conjunction with the end users. While I regret that the digging and cultivation garden did not work out for its users, it is part and parcel of the way people adapt the setting to their own needs, to make an ever-improving environment for play. Looking at it now, it feels even better than when it was first completed and handed over. The climbing frame has added drama to the third room, and in its totality this small space is so full of drama that children often talk together about their experiences using it. This is as it should be.

Some conclusions on play in an adult world

There is a popular contemporary debate which seeks to set the historical or evolutionary roots of the human species apart from nature and instead to suggest that since the advent of the age of technology at the end of the 1960s we are all simply the outcome of our own genius. One can argue that as the phenomena of global warming becomes more real, standing apart from nature can only be viewed as a short-term selfish ideal. If as architects we design a children's building with only the final fee as the motivating principle, any scope for considering broader, more holistic principles is just about impossible. So it is right to ask ourselves, what is the nursery for, at its most basic level?

Many experts now consider the development of the human brain to be the primary aim of the exercise. As Aric Sigman states[16], 'between the ages nought and three, particularly when children are acquiring language skills, their brains are going through rapid development and are being physically shaped, like a piece of clay in response to what they are exposed to'. Therefore the environments to which we are most exposed as children, the nursery and the home, should be considered in those careful terms. The reality of nourishing sensory perception is arguably the single most important aspect of the designed environment, both inside and outside. This will help the child's brain to grow in the most healthy ways.

In a simplistic sort of way, it's all about poetic delight within the environment. Until there is a greater awareness of the importance of this exposure to sensory stimulants with a high aesthetic dimension, sadly as

The natural child

educators we must develop environments that have value which is merely quantifiable. The learning environment guidance which states that 'intrinsically linked to sensory experience is emotion.... Children need to have experiences which heighten emotions such as wonder, joy and excitement, and children need adults who will bring out and develop these emotions' is not saying that the children's garden should be equipped with a few milk crates and a plastic half-round gutter for children to savour the delights of water play.[17] Yet this prosaic image is used on the very next page after this marvellous quotation of a guide to planning the perfect outdoor environment. Do they not realize that children understand the quality of a space and how this affects their development and sense of self-esteem? We need to be much more in tune with nuances of play and how children, rather than adults, relate to it.

It is evident that every sensory perception is partly governed by biological memory, and partly determined by the cultural environment the child is exposed to on an everyday level. If children experience a poor cultural environment where tactile and olfactory perceptions are restricted, just think how their nursery environment could step in and fill the cultural gap. Outdoor play spaces, in particular, can potentially offer valuable experiences of the outdoors if designed with sensory values at the fore. They can provide contact with living things like plants and animals, and promote an awareness of environmental conditions which change with the seasons. This contact can enhance physical and cognitive development, encourage imaginative play and stimulate empathy. Some experts even attribute restorative qualities for damaged or disruptive children. The power of the senses should not be underestimated.

The *7Cs* researchers found, when they asked early childhood educators what they wanted to see in their play space, 43 per cent would like additional sensory experiences, 35 per cent would like to see better organized space, 22 per cent would like better equipment, structures and seating, and children themselves wanted more soft spaces in both their inside and outside areas. There is in this slightly bland summary a sense of incredible richness which touches upon the power of good landscape and architectural design in outdoors spaces. As designers, we should view ourselves as terrorists, planting little time bombs of thoughtful, complex play spaces which are full of sensory delight. Usually we do this despite the staff who may object to the whole notion of complexity in childhood play. Narrow minds need expanding.

CHECKLIST 3: KEY DESIGN GUIDELINES FOR THE OUTDOOR AREAS.

1 Secure storage outdoors which is accessible to children works on two important levels: it eases pressure on limited space inside the building; and it provides extended play and interesting learning experiences outdoors.

2 A cover for sand-pits when they are not in use will keep cats and foxes at bay.

3 Create clear pathways around the garden, but consider also the idea of mystery by using low walls and planting, kinks and turns in some of the routes to make the pathways more interesting and to slightly obscure the journey for children without compromising adult supervision. By doing this children will be far more engaged with their play.

4 External trellis areas covered in climbing plants create important meeting points or dens for children; different-sized areas give various options for children to hang out in small or large friendship groups.

5 Encourage cultivation activity within the curriculum with dedicated flower, herb and vegetable patches. Encourage parents to get involved: create planting beds for exploration; children's seating areas should be located around the garden; a separate fenced-off area for babies should be provided (if there is space), so they can be safely outside at the same time as older children; provide areas for climbing and running if there is space; construction areas for the making and enjoying of building; a sand and water area has tremendous learning benefits; recognize and communicate the experience of the changing seasons in your planting plans; there are flowering winter plants as well as those for high summer; autumn is a time for lots of supervised work in the garden; a sensory garden can be created where children can recognize objects by sight, sound, touch, smell and taste – but only if there is space.

6 Take care of the outside environment – treat it with respect and concern, and show this to the children to build a higher sense of value. Young people's basic need for well-being and involvement and their urge to explore and make a sense of the natural world is developed through high-quality play in an outdoor environment; avoid synthetic outside spaces – there is no such thing as a maintenance-free garden.

7 The indoors and the outdoors should be seen as one integrated zone with easy run-in–run-out play.

8 Children need some control to change and modify their environment: sand, leaves, water, mud and soil are great manipulative components; large blocks, logs and wooden planks can also be used by children to modify the space.

9 Consider gender issues in young children's play and if there is space consider providing quiet protected areas for contemplative play and large physical equipment for physical play.

10 Like any garden, nursery gardens need maintenance to make the most of what can be a truly transformational experience for children. Consider asking staff and parents to take responsibility for individual parts of the space; form a garden committee.

Chapter 4

A historical overview

Form becomes feeling

Background: establishing principles

The history of childhood and early childhood education is long and complex. It is argued about up to the present time, with important educationalists claiming their doctrines to have been the first or the most significant. If radical thought is at the very root of innovative ideas in education, then experimental pre-school systems have perhaps been the most radical, initiating ideas like 'child-centred learning', 'open teaching' and the 'outdoor classroom', which anticipated many later developments in general education. A number of landmark buildings reflected these progressive ideas. It is fair to say that they have on various levels acted as prototypes for progressive school design, as well as early-years education up to the present day.

  The aim of this chapter is to set out, in the briefest form, these key ideas as reflected in the visionary individuals and the environments that followed. This forms my own somewhat potted history of nursery architecture. I wish to describe some of the key buildings that illustrate the ideals in action, and reflect upon the more recent late-twentieth-century examples of architecture for early years which provide an ethical template for the next generation of nursery school architecture yet to be built. We also explain how some of the key educational pioneers managed to bring about a convergence of progressive educational and architectural thinking, thus creating environments which could, and often succeeded in, transforming the lives of children for the better.

  It is important to consider the nursery or kindergarten within the framework of a generally progressive and longstanding tradition based on two key principles. First, child welfare or protection with professional and hopefully loving care within a secure building; and second, accelerated child development in particular through the catalyst of an environment which supports learning through play. It was and remains a socialist concept based on fairness and equality of opportunity for the most needy in society. However, often in those early years it was adopted by middle-class visionaries who recognized the benefits which came to their own children. The view that the child's earliest years are the most critical in defining his or her future success is now more widely understood.

Nursery or kindergarten has only recently been associated with commercial enterprise and the privatization of some early-years services, with young children being viewed in some quarters as 'units of profit'. During the course of an interview for an architectural job with one of the largest UK nursery chains, I was somewhat perplexed to hear the managing director's comments in relation to his work; comparing it to the airline business, he explained that losing a child while in care was similar to losing an aeroplane, it inevitably meant the end of the business. This voracious and uncaring attitude is all too common within the privatized nursery sector. The idea that you can make money out of childcare is a largely UK and US concept; after all, if profits are squeezed, you can't cut down on the staff numbers to save money as would be the case in most other commercial enterprises, whereas investing in good-quality design is one area where costs can be limited and constrained.

Having said that, the awareness of good design for early years in the public sector is still 'hit and miss'. While central government pay lip service to the notion of high-quality early-years services, clear definitions as to what constitutes good quality are hard to pin down which, given the importance, is puzzling. Following a recent survey of Sure Start facilities in England, it would appear that design of the highest quality is evident in only 20 per cent of public sector early-years buildings. The 20 per cent architecture – with the rest just building – may I guess be applied across the spectrum of built work, certainly in the United Kingdom and United States; however, in early years poor-quality environments have a more profoundly negative effect on the development of its primary users, the children, than is the case for most buildings for adults. Children are simply more environmentally aware. Unfortunately, many Local Authority commissioning bodies are more interested in low-costs, especially at this time of squeezed public spending.

As much as this might be a financial issue it is also a cultural problem in the United Kingdom, and I suspect in the United States as well, where architecture and good design (as far as that might be defined), is still widely viewed as being irrelevant and costly. 'We are not designing an art gallery' I hear them say. 'Outcomes' are not to do with the real needs of children, who I am convinced appreciate good design and thrive on its nurturing ambience, 'outcomes' in this profit-obsessed world are more about efficiency in the means of production, how many children can be serviced on a minimum-cost base. A slightly cynical view maybe, but borne out by long years of experience, and it is worth stating in the context of this brief history of nursery architecture, it was ever thus.

The point I make here is that although a number of examples of progressive nursery architecture are cited here, they are iconic rather than typical. Nevertheless, they help to define the principles upon which good contemporary design for early years can be checked and quality controlled, which hopefully takes us beyond the normal prosaic checklist of rooms for functions and layouts for control, a staple of any briefing document for architects. What a low-level set of aspirations! Nursery is about much much more; children's rights in the broadest, most humanitarian sense means care and nurturing, but beyond that the constellation of sensory perceptions which the best environments can stimulate so that children can make sense and progress within their world. And beyond that, everyone involved having the courage to create spaces which challenge children to test their physical and mental boundaries while being secure and protected. These qualities are hard to value in this commercial world, which has been created for us and by us.

One of the key historical themes which emerges time and time again in this text is the critical transformation in the lives of children as a result of the Industrial Revolution. Much of this was down to a fundamental disconnection between nature and the everyday lives of children, with the mass migration from the land to the newly industrialized cities. This became more and more apparent as the seventeenth and eighteenth centuries evolved from their highly communal rural form towards a more atomized social structure in the urban milieu. This dissolved important social structures which had previously provided family support in natural, organic ways.

However, this transformation also showed itself in other areas, such as the domination of industrially manufactured materials in building design, particularly since the Second World War. This meant that the environment was smoothed out and its materiality made expressionless for largely economic reasons. As a result, tactile qualities of the more traditional environment were denied to children. Smooth, synthetic, often chemically derived plastics and metals, processed and industrially manufactured to within an inch of their being dominated. They largely took the place of roughly textured natural materials one step removed from nature, thus denying the affordances that so enhance child development during the early years.

The industrial process even promoted some materials which were known to be carcinogenic. If spray-painted MDF replaced a hunk of unplaned oak in the construction of children's furniture, and plastic floor tiles were used in place of a length of rough-hewn stone, the effect on children's sensory understanding would be a little like a blind man trying to relate to an environment where all the other senses such as smell and hearing were absent. In terms of the child's heightened perceptions, what we end up with is a place which does not really exist. There is simply 'no there there' and it ceases to be a truly child-friendly environment.

Some fear that this distance from nature is even more marked since the advent of digital technology. Writing in 1992, architect Peter Eisenman states that since the Second World War, a profound change has taken place in the ways in which we interact with the world. He describes this process, in somewhat jargonistic terminology, as 'the electronic paradigm'.[1] This alludes to the shift from mechanical to electronic devices which, he stated, would increasingly dominate our lives; in this he included television, fax machines and photocopiers. What he did not predict was arguably the most profound social transformation since the Industrial Revolution, the advent of user-friendly personal computers, the internet and the World Wide Web.

This has undoubtedly affected the lives of children more profoundly, providing an easy escape from the physical and multisensory challenges of the natural world, which were the ready-made playground for previous generations of children. As must be clear to the reader, I believe that a closeness to nature is fundamental when designing a nursery environment. Yet today the majority of nursery providers emphasize the benefits of computer activities for the youngest people during their most formative years, rather than restricting their use; there appears to be a lazy consensus that the sooner they become computer literate, the better. No-one questions the need for more holistic activities within the nursery, instead (one suspects), sophisticated visual distractions on screen are viewed as an easy way to keep children quiet – Ritalin in digital form.

Children's rights to a more natural environment are denied to them so categorically in many of the second-rate nursery buildings I see everywhere I go, that one wonders how many of these places are ever passed as being suitable for use by children. Many childcare professionals simply fail to recognize the importance of environmental complexity in raising the game for child development. Fortunately there are some exceptions to counter this bleak view, and these are the projects celebrated here and elsewhere in the text. Whenever considerations for essential child perceptions are asserted within the design process – if you like, the channels through which they navigate their world – when these faculties are considered deeply and carefully, children's rights are given a fundamental priority which makes their lives in care positive rather than negative for them at this most crucial time.

There are a number of key educational innovators who first recognized the importance of children's environments. Although opinions differ, the kindergarten/nursery idea is generally considered to have been initiated as early as the seventeenth century, when small groups of people in a number of European countries became interested in the moral reform of society. Churchmen, lawyers and scholars recognized the importance of education in bringing this about. At this time, Comenius talked about organized childcare. In the eighteenth century, Rousseau rebelled against the pervasive moral atmosphere of pre-revolutionary Paris, asserting the beneficial effects on

David Stow's first model infant school, Drygate, Glasgow (1828). The three key institutions of the state are shown: the school, the church and, on the hill, the factory. Note the school master joining in with childhood activities, complete with his tophat. Reproduced by permission of Jordanhill College of Education, Glasgow.

A historical overview

children of an outdoor life. We will touch on these early influences lightly, but for those interested in a deeper historical understanding of architecture for early years, it is an area worthy of further research.

It was David Stow (1793–1864) who first came up with the idea of the nursery as a sort of nirvana when, in the context of his first Glasgow Infant School (1828), he stated that 'The playground is a Garden of Eden', drawing obvious analogies between the biblical story of idyllic innocence prior to the first and the original sin.[2] It was his contemporary, the pioneering educationalist Friederich Froebel (1782–1852), who almost certainly came up with the term 'kindergarten', with its somewhat biblical connotations relating children to the idea of budding flowers nurtured like delicate seedlings within the warming environment of the nursery. This powerful image provides the ethical template for most good nursery architecture up to this time.

Johann Heinrich Pestallozi (1746–1827) was an influential Swiss pedagogue and educational reformer who exemplified Romanticism in his approach to child welfare. He published his masterpiece, *Leonard and Gertrude*, in 1781[3], an account of the gradual reform of an entire German village by the efforts of a good and devoted woman. In 1801 he explained that his educational methods were based on a process which went gradually from easy to more difficult tasks for the child. He emphasized the enlightened view that every aspect of the child's life contributed to the formation of personality, character and reason. He even promoted the optimistic idea that human nature was essentially good. He believed that there were distinct phases in the educational process, starting with observation, or learning to see, which was followed by consciousness, and then from consciousness speech began to form. Then came measuring, drawing, writing and reckoning, the key constituents of a conventional education. In 1887 he established a school at Burgdorf where he experimented along these lines.

He was of course reflecting upon the age-old quandary, how do you represent ideas to young (pre-verbal) children? The importance of drawing in the development process was asserted by Pestallozi perhaps for the first time. Pedagogical drawing was distinguished from academy teaching in that it began at an early age and its exercises could be promoted to groups of pre-literate children within the new German kindergartens. The key theory that counting, measuring and speaking could be coordinated by drawing was, in practical terms, about identifying individual forms, analysing their shape and using language to describe them. His influential publication *ABC der Anschauung*, written with the assistance of Cristoph Buss in 1803[4], helped to establish drawing as an important component of the educational process for every child. Today it seems obvious that arts and crafts are essential curriculum activities; however, prior to Pestallozi it was hardly recognized.

Froebel was without doubt the most influential early theorist. His boyhood interest was in the natural world and initially he undertook an apprenticeship in forestry. From 1800 he studied biology and mathematics at the University of Jena, where he first heard of Schiller's part in the formation of Naturphilosophie. Imbued with an interest in the sciences balanced by his newly found philosophical creed centred on nature and its phenomenological importance, he went to Frankfurt to commence a course in architecture. Although he did not complete it, the skills he acquired were to be put to later use in the design of his own houses and school buildings. Having been invited to teach drawing at a school in Frankfurt, his teaching vocation gradually emerged.

From 1807 to 1810 Froebel worked under educationalist Johann Pestalozzi at his school in Yverdun. Froebel recognized the importance of creative development through play as opposed to unrelenting discipline, which underpinned much mainstream educational theory of that time. He began to see how important it was to cultivate the uniqueness and individuality of each child. He believed that children had an almost mystical understanding of the innate truths of life, and that this spirit could be reawakened by playing games which had symbolic meaning, always referencing nature. For him, the kindergarten should represent an ideal society – hence its name. This referred to the environment in its entirety, including the gardens and buildings which, designed together, could take on all of the representative symbols of the biblical Garden of Eden. This was important since he believed that young children understood through a symbolic, almost mystical language which at its most informative utilized metaphor and analogy, often derived from natural phenomena and/or mythical stories.

One can see this in contemporaneous nursery rhymes such as *Ring-a-Ring-o'-Roses*, an allusion to the healing powers of natural remedies during the Black Death, or *Grimm's Fairytales* (first published in 1812[5]), such as *Little Red Riding Hood* and *Hansel and Gretel*, both allusions to the safe world of the village as opposed to the dangers of the forest/wilderness. However, there were other underlying messages about the relationships between children and close family; for example, in the first version of *Sleeping Beauty*, the wicked step-mother was originally the mother of the heroine, perhaps reflecting the underlying Froebelian message that the kindergarten was a safe haven for children from the perils of the world, many of which were to be found in the home – a cautionary message which is readily transferable to modern times. However, the publication of *Grimm's Fairytales* and the work of Froebel were also clear indications that in Germany a recognizable children's culture was beginning to emerge.

Having visited a number of so-called progressive nursery schools, which he viewed as no more than child storehouses for the convenience of working parents, in 1837 Froebel established his own school. This he called a 'school for the psychological training of small children through a system of play and occupations'. As an aside, it's hardly a catchy title for an early-years institution, but perhaps slightly better than more contemporary nomenclature such as 'Teddies' or the risible 'Little Darlings Nursery', both in London. What he meant was that the traditional role of the active teacher and the passive child learning by rote would be reversed; instead children would be given a wide range of materials and encouraged to carry out various sorts of creative and expressive handwork, thus self-activity became the means of education. He believed there were forces within children which move them towards those activities that stimulate development, and if they were correctly trained the restless activity of children should be sustained and directed by the teacher towards these developmental goals. This was a truly revolutionary concept for that time.

Gradually, Froebel became the most influential early pioneer of a new, enlightened form of childcare. Froebel believed that young children

A historical overview

benefited most by structured free play in school and at home, as opposed to restrictive discipline intended to 'break the will of the child'. He devised one of the first educational toys and called this revolutionary invention 'the gifts and occupations'. It featured natural wooden building blocks, which encouraged close hand–eye coordination. Their shape and form were derived from natural geometry, which Froebel believed aided spatial comprehension, and many toys based on these principles are still widely used today, including basic building blocks, posting boxes with different shaped holes and wooden jigsaws. Indeed, the Froebel Company, based in Grand Rapids, Michigan still manufacture the gifts and occupations in their original form. It is worth pointing out that my own children have used the blocks throughout their early years. During the years of their childhood the blocks have proved to be an engaging and pleasurable alternative to other pre-bedtime activities. Their benefit is as much in engaging the parent as it is in engaging the child.

The first kindergarten based on Froebel's ideas in Britain opened in London's Tavistock Square in 1840 (an area which is the centre of child welfare research to this day). It was an environment that seemed to focus almost exclusively on the garden rather than the building. Later, the Ronge sisters – who had originally come under the influence of Friederich Froebel in Germany in 1818 – were responsible for the first American childcare facilities. They further developed the games children would play while in their care, let alone play with toys. Froebel emphasized that the physical activities and restlessness of the child should be sustained and directed towards developmental goals rather than being crushed by a sense of overbearing discipline. He and his co-teachers provided a wide range of materials for creative play, and encouraged children to carry out various sorts of artistic and expressive handwork: self-directed activity became the means of education, and for Froebel, a correctly designed environment became a crucial catalyst towards these goals.

Froebel's learning system is said to have profoundly influenced the work of many architects in its three-dimensional manipulation of space and form. Prior to Froebel, it would have been unusual to suggest that children would play with geometric building blocks as part of their education. But he provided a framework within which later early-years educators could structure their educational systems, not just toys and activities, but also the spaces within which structured play would take place. The great American architect, Frank Lloyd Wright, was said to have been profoundly affected as a child by his activities in a Frobelian kindergarten. He designed buildings which were in harmony with the educational ethos of Froebel and his naturalistic philosophy.

Frank Lloyd Wright's influence at the hands of his kindergarten teachers is more than just academic. It is worth describing in a little more detail since arguably his architectural legacy remains profound to this day, and this we will explore further on in this chapter. The youthful Wright explained how he and his mother worked together with the Froebel 'gifts', which became the source of profound pleasure and his subconscious awakening to the primacy of shape, texture and form. He describes his engagement with the Froebel block system as follows: 'The smoothly shaped maple blocks with which to build, the sense of which never afterwards leaves the fingers: form becomes feeling.'[6] This influence is clear to see in his subsequent architectural genius.

A historical overview

Another early pioneer was Robert Owen (1771–1858), who knew of Froebel's work and had visited his experimental school at Yverdun. At the time, no-one was better known as a pioneer of education for young children in Britain than philanthropist Owen. In 1816 he had opened the first infant school in the country at his New Lanark cotton mills. He was impressed by the possibilities of industrialized production, and dismissed notions of the revival of a pastoral society. He foresaw that new sources of wealth might be used as a socialist tool for the benefit of working people, and also recognized the problems the new system might bring to the lives of young children, with long working hours for parents and little opportunity for good progressive care for their children. Most importantly, he believed that human character was formed by the individual's environment (seen in its widest sense), rather than being imposed upon them by an individual disciplinarian teacher, preacher or parent.

Owen devised three schools at New Lanark. Children between the ages of two and six years attended the infant school, which was the one in which he had the most interest. Books were excluded, and activities such as singing, dancing, marching and basic geography took their place. The children spent three hours per day of free play in an open playground, unless the weather was bad. Initially the school combined nursery and infant activities; later, separate rooms were used for the 2–4-year-olds and the 4–6-year-olds. At the age of six or seven, children moved on to the schoolroom. At ten, they left to work in the mill and attended evening classes with the adults.

Owen established his new school as part of his model factory settlement at New Lanark in the Scottish Borders and called it 'The New Institution for the Formation of Character'. The school was opened in its basic form, as a two-storey brick-built shed; the lower storey for the younger children is almost open plan, divided into three apartments of nearly equal dimensions, 12 feet high, with the wide spans supported by hollow iron pillars, serving at the same time as ducts for the distribution of warm air from a centralized heating system. The space was enlivened by the positioning of framed

Interior of Robert Owen's New Institution for the Formation of Character (1880). Reproduced by permission of the New Lanark Trust (www.newlanark.org).

naturalistic objects, pictures of wild animals and other sensory stimulants; this was another innovative idea, but in line with Froebel. Owen even sanctioned the purchase of a live alligator as a valuable hands-on stimulus for the children. However, it was an experiment that did not last very long for obvious health and safety issues. Besides, the alligator perished as soon as winter set in.

The very act of providing an alternative to children being in the factories (beside mother) improved the cotton mill's output significantly and eradicated many child-related accidents. However, there is no doubting the importance of the building and its enlightened nurturing approach within the overall structure of the system. Owen's school cared for children from the age of 18 months to ten years. Infants aged from two to five spent only half their time in the school – during the afternoons they were allowed free play in a large paved area surrounded by plants and trees. In the mornings they enjoyed active play in one of two lofty play halls which were well ventilated and full of light:

> They were trained to habits of order and cleanliness; they were taught to abstain from quarrels, to be kind to each other. They were amused with childish games, and with stories suited to their capacity. Two large airy rooms were set apart, one for those under four years, and one for those from four to six. This last room was furnished with paintings, chiefly of animals and a few maps. It was also supplied with natural objects from the gardens, fields and woods. These suggested themes for conversation, and brief familiar lectures; but there was nothing formal, no tasks to be learned, no readings from books.[7]

As implied by this quotation, the approach to care was quite advanced pedagogically, using the principle of largely self-directed play whenever possible, although there were a number of group activities on offer, as can be seen in the somewhat idealized internal perspective – there was dancing for the girls and marching for the boys. Children were not made to participate in the activities if they did not wish to, and sleep sessions occurred

Froebel teachers Carrie and Tyra Wiltheis of the Wiltheis Kindergarten School, Seattle, watch over the children in class who appear to have built different forms of model factory buildings.

The Froebel gifts that are available to buy in the United Kingdom from Education Design Toys.

A historical overview

Amy Dudek plays with the Froebel gifts, numbers 5 and 6. With direction from her on-looking dad, she is given guidance on how to make the individual blocks into larger whole units, and, most importantly, how to replace the blocks in their boxes when she is finished. The level of concentration on her tasks is significant for a three-year-old.

The author uses gift number 6 to construct a version of the Avery Coonley Playhouse façade. With its strongly layered ordering and blockish style, it could almost have been designed in model form. Frank Lloyd Wright stated of his childhood play with the Froebel blocks: 'The smooth shapely maple blocks with which to build, the sense of which never afterward leaves the fingers: form becoming feeling'. (Wright, F.L., *Frank Lloyd Wright: An Autobiography*, Duel, Sloan and Pearce, 1943, pp. 13–14).

Frank Lloyd Wright's Avery Coonley Playhouse was commissioned in 1912 by Mrs Queene Ferry Coonley for her children. For a time she also operated a small school there. Wright's delightful designs for the windows match the purpose of the building itself, complementing its modernist abstract style. From the windows to the furniture and fittings, this is a building which is truly in-tune with childhood scales and perceptions as Wright understood them; his architectural inspiration undoubtedly comes from the geometric proportions of the Froebelian gifts and occupations he used as a child.

whenever the individual child wished. Where the Owen system differed from the Pestalozzi and Froebelian approach was the notable absence of arts and crafts activities: the emphasis was on the physical rather than the intellectual and emotional development of the child. Nevertheless, the sound originality of Owen's curriculum has been praised, but clearly his methods were felt to be deficient in the more imaginative and creative aspects. Owen himself was not interested in literature, although he was emphatic about the importance of speech and reading – however, for Owen this would be largely without poetic content. There were simple lessons in drawing for the top classes, but no painting, craft activities or creative play activities at all were mentioned in connection with the nursery school curriculum.

Pestalozzi, Froebel and Owen made significant contributions to nursery school education in these early times. Yverdun became something of a laboratory of early nursery school practice, visited by a number of English admirers who then returned home suitably inspired to open their own nursery schools in the shadow of the Industrial Revolution. Owen could be described as being at the more practical Anglo-Saxon end of nursery practice. His importance in the context of this study is that his was the first recorded institution where the environment was a primary consideration in that it was purpose-made, and it can be assessed on the basis of the engravings and buildings which still survive. Although architecturally primitive by today's standards, it was a building designed specifically for the needs of young children and perhaps the first recorded example of an institutionalized architecture for young children.

Margaret McMillan, a much-respected British pioneer of nursery education, alluded to the ideal nursery school in 1926 as a 'garden city of children'. Her school in Deptford, East London might be described as a progressive lightweight glassy construction and comprised a number of small, independent pavilions standing in the school garden. Each housed a family unit of 14 children with two nursery nurses. It had its own bathroom and its own play equipment designed to meet the needs of a particular age group. Its intimate home-from-home scales provided an ideal place for small children, and it became a template for subsequent developments in kindergarten architecture.

McMillan did not believe that the nursery should be like a conventional school; rather, she promoted the idea that it should have purpose-designed buildings relating to the age of the children, each one housing a family-sized group of children with its own mini kitchen, children's washroom and the usual activity zones, for play (with floor-level activities) and rest.

While it is not always possible to follow her ideas of pavilions set within a garden (because her proposal is costly compared to a more unified plan form), a number of her design themes still have relevance for us today, particularly the idea that the nursery could help to replicate a typical family environment. For example, the idea that indoor activity spaces should open directly onto the gardens to provide run-in–run-out play for children is tremendously important to her initial thinking and subsequent developments. Her view about the buildings themselves can be interpreted as the need to provide interiors which have a soft, child-oriented quality, with carpets or rugs on the floors and drapes and soft fabrics on the walls. Softness as opposed to

A historical overview

Margaret McMillan, the open-air school in Deptford, with children taking a morning nap. Reproduced by permission of Lewisham Local Studies and Archives. A plan of Tarnerland Margaret McMillan Open Air Nursery School, Brighton, which was inspired by a visit Margaret McMillan made to Brighton in 1930 to lecture members of the new nursery school committee.

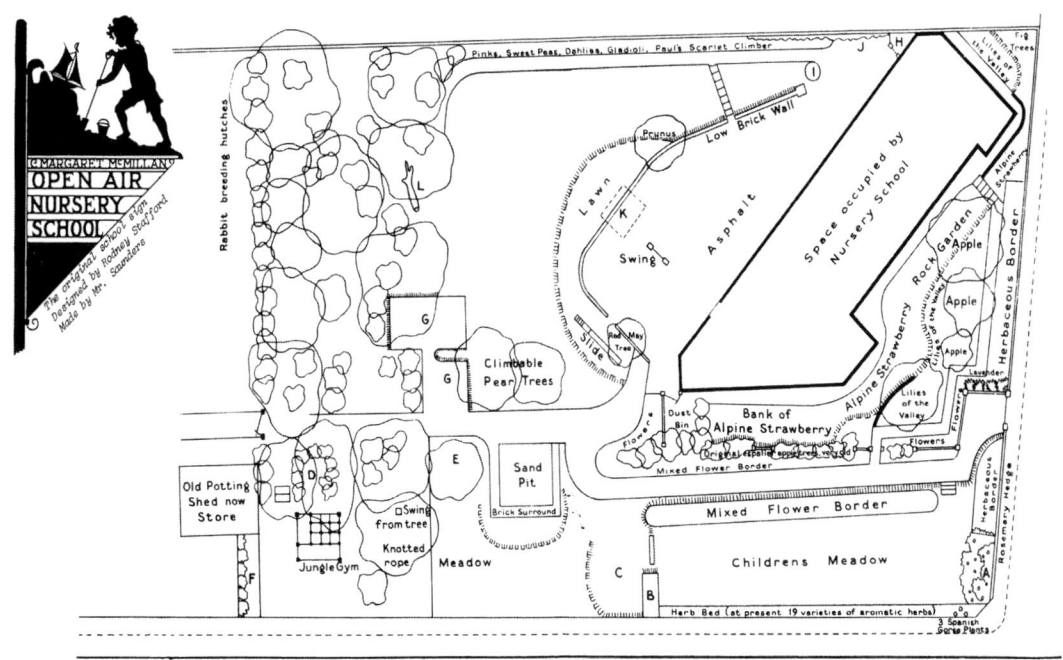

A. Plum suckers pruned to form play caves.
B. Play shelter made from an old door with turf walls.
C. Place for excavation of all kinds.
D. Play house formed by euonymus shrubs.
E. Old drooping apple tree that makes a play house.
F. Lavender hedge.
G. Specially levelled ground to take dinner tables in summer.
H. Fence to prevent children in the garden approaching the isolation room window.
I. Old ship's capstan used as a bird table.
K. Proposed site of outdoor brick built Wendy House
L. Fallen tree trunk placed against growing tree for climbing.

THE GARDEN SLOPES IN ALL DIRECTIONS FROM A WHICH IS THE HIGHEST POINT.

hardness, small-scale as opposed to being over-sized and institutional, with an emphasis on the garden rather than the confined interior; these were the three key principles. Most interesting of all was the idea that each school-home (or in modern parlance, the home base) would have its own specific day organized especially to provide a ready-made curriculum structure which

A historical overview

replicated the needs and cyclical patterns of the home. It would be particular to the children using it, rather than a standard day arbitrarily divided into 45-minute lesson periods as in the school. Her perceptions are encapsulated well in the following quotation:

> The idea of a large and strongly built edifice as a school for children went by the board long ago. To hold such a conception (and it was long held), is as if one, escaping from a cave-dwelling, insisted on living in a large prison ... the school of tomorrow will be a garden city of children; that is to say a place of many shelters – a township, if you will, of small schools built as one community but with every shelter organized as a separate unit designed to meet the needs of children of specific age or stage of life.... Every shelter is in effect a small school. It is also a self-contained unit or schoolhome; it has its own Head; it also has its own bathroom, its own equipment, and its own school day, adapted to the needs of children at a specific stage of development.[8]

This of course was pioneering primarily because it initiated the idea of the open-air nursery, a concept which has since evolved to suggest varying levels of interaction between the inside and the outside on the basis of its health-giving benefits. It underpins profoundly the entire mainstream nursery ethos of the twentieth century. Writer and historian Andrew Saint, describing post-war school buildings in England, firmly acknowledged the debt to Margaret McMillan and the nursery school movement and pointed out how in the 1950s and 1960s 'this movement became a channel through which "child centred thinking" could filter into local authority teaching'.[9] Nursery schools had single-storey layouts, flexible schedules, low budgets, short days and were full of light and air. The new primary schools mimicked many of these principles, stretching the play-based ethos up into the activities of 5–8-year-olds in primary school.

Margaret McMillan had a passionate, romantic view of childhood, and believed that the lives of young children in slum districts could be transformed if they attended open-air nursery schools. As a health educator and socialist crusader, she had a clear view of what nursery schools could do for poor children. They should offer all-day care, from 8 a.m. to 5.30 p.m. or 6 p.m., with regular meals – breakfast, lunch and tea – and a long nap after lunch. In the poorest districts they should provide residential overnight accommodation, where possible in the open air. The garden was above all the distinguishing feature of the new type of school:

> All the best apparatus is in the garden.... Out in the garden we have tried in spite of our difficulties, to plan generously, remembering that to deficiency of material we own perhaps most of our failures. We have sown or planted well over a hundred kinds of flower and blossoming shrubs and at least twenty orders of tree ... neither do we put flowers in spots or pots: the first impressions should be massive so we have great breadths of purple, blue, crimson and gold in our garden.[10]

When state education was first introduced in 1868, the age of five was arbitrarily decided on as school-starting age (as opposed to six or seven as in many European countries). Even so, many children under five went to board schools. In 1900, 43 per cent of 3–4-year-olds were in schools, a much higher percentage than today. The schools were used as substitute childcare by working parents, younger children accompanying older brothers and sisters to school. For the schools, then as now, the incentive was to increase numbers, and prepare children as early as possible to sit competitive examinations. But there was increasing unease about the suitability of the school environment for very young children. The classes were very large, and the children had to sit still and learn by rote. McMillan objected to this.

Although she was very well known in education circles, and president of the Nursery School Association, her view of nursery education was unconventional, and she was at war with many of her co-educationalists from the outset. Her notion of nursery education arose out of her concern for the health and welfare of poor children rather than stressing their educational progress while attending. Her stress on fresh air and aesthetics was in order to rescue such children from what she saw as ugly, unhealthy and dispiriting slum conditions. Hence also her emphasis on daycare and even residential care as an important aspect of nursery education.

Many of her colleagues in the Nursery Schools Association had come from different, more middle-class routes. They had been influenced not by welfare socialism, but by the ideas of Pestalozzi, Froebel and Montessori. These educationalists put forward ideas about children learning through their experience, through active imaginative play, and carefully constructed play environments. Their ideas were taken up in Britain from the end of the nineteenth century – there were kindergarten societies and training colleges for kindergarten teachers and an increasing number of private nurseries emerging at this time.

Some of the key kindergarten ideas were developed by Susan Isaacs (1885–1948) at her demonstration Malting House School. She stressed that 'mothers and nurses have begun to turn away from mere custom and blind tradition to science'.[11] She considered that there was a right and scientific way to bring up young children, to allow them to play freely in carefully designed spaces. She had a champion in Bertrand Russell, who in a best-selling pamphlet wrote: 'On Education especially in Early Childhood even the best parents would do well to send their children to a suitable [nursery school] from the age of two onwards at least for part of the day.'[12]

In the 1950s the idea of learning through freely chosen play in a structured environment got further theoretical boost from the developmental psychologist Jean Piaget. He postulated a theory of learning in which a young child continuously experimented with her [sic] environment and built up her own theories about how the world worked; the child as a self-motivated empirical scientist was a strong and abiding image and should by right have brought the architecture into focus more clearly as part of this 'laboratory' concept. However, at a time that 'functionalism' derived from the pragmatic needs of people was all the rage, little was specifically made about the architectural theories that might support the children's activities. Piaget remained very important up until the 1980s, when his preconceived-stage ideas were questioned.[13]

Unlike Piaget, who conducted observational research in his clinic, Susan Isaacs recognized that children are much more likely to show their true nature in environments like the Malting House School, where everything was centred on a large garden for free play, which included a chemistry laboratory and a woodworking area. Susan Isaacs was a Lancashire-born educational psychologist who stated clearly that children learned through play and that it should be viewed as children's work. She said: 'What imaginative play does in the first place is to create practical situations which may then often be pursued for their own sake and this leads on to actual discovery or to verbal judgement and reasoning.'[14]

She also pioneered the method of naturalistic observation, documenting the children's experiences in great detail and articulating their activities in a refreshing style that was devoid of sentimentality but full of life and wry observation. Her's was not a protective, safety-conscious view of childhood; rather, she saw children as scientists with a unique view of the world, constantly experimenting and taking risks with their limited knowledge of the world. The following story was told by Mary Field, who produced a film in the 1930s of the Malting House School. On visiting the school for the first time, she got the impression that:

> some of the activities had been laid on specially for us. For example, the children were dissecting Susan Isaac's cat, when normally they worked with frogs or dogfish. They all seemed to be enjoying themselves immensely, digging away at the carcass.... Then there was the bonfire. It was supposed to be an exercise in free play, but it got a bit out of hand. The fire spread and spread and reached the apple trees, and then destroyed a very nice boat. Even Geoffrey Pike [the owner], was a little upset about that, and he seemed a very calm man.[15]

During the 1950s a further theoretical influence of a different kind came through the teachings of the psychoanalyst John Bowlby. He held that separation of a child from his mother was harmful before the age of three, and only slightly less so over three if the separation was for more than part of a day. This theory took deep hold in England and a series of circulars from the then Ministry of Education advised Local Authorities strongly in favour of part-time education and part-time care. McMillan's beds on the veranda concept finally became obsolete as full-time daycare was phased out throughout the 1960s and 1970s; all but the most vulnerable or at-risk children attending full daycare nurseries (which were provided by social services rather than education) became a thing of the past. This tendency continued in the United Kingdom right up to the 1990s, when private nurseries took-up increased demand from working parents for full daycare.

Over the final decade of the nineteenth century the kindergarten idea had become increasingly institutionalized and somewhat frozen in time. The conceptual basis of the entire movement became fixed within the framework of Froebel's dated methods. Radical change was inevitable, as new answers to new problems were identified. The world sought a more progressive basis for early-years practice, moving away from Froebel's vaguely mystical musings towards a more scientific approach that reflected the advent of High Modernism in art, architecture and design. At this time Maria Montessori's (1870–1952) influence came into fashion. She maintained some of his ideas while introducing a range of radical new child-centric concepts, which she had discovered in her medical practice, particularly in relation to her work with children who had intellectual disabilities. Published in *The Montessori Method* (1915)[16], it was a universal philosophy which had a much more rational, pragmatic basis looking forwards rather than backwards, emphasizing physical and mental coordination with revolutionary new equipment to facilitate these processes; for the Montessorian child, 'the hand was an extension of the brain' and if provided with the right apparatus or exercises as she called them, children would self-direct and self-teach.[17]

Montessori was a medical psychologist who became concerned about the living conditions of young children in Rome at the turn of the century. In 1906 she became the main organizer of the new infant schools throughout Rome. She recognized that for young children it was not a general curriculum that was required, rather it was the need to address the particular needs of each child. She recognized an interrelationship between mental and physical capabilities and developed play structures which combined both. In its most basic form, a staircase would, if designed with the child in mind, become a feature with immense developmental value. For Montessori, architecture and the built environment in general held similar value for the young child and her carer. One of the first purpose-designed Montessori nursery buildings, the Haus der Kinder in Vienna of 1926, was designed by Schuster, and was an exemplar of its time.

Montessori began her research using the basic idea of scientific education, an idea which had been developed in the 1800s by French physicians Jean Itard and Edouard Seguin for children with special needs. This centred upon the initial idea of observing children at play, encouraging what they viewed as their natural, free activity by adding a series of exercises with specially designed self-teaching materials. A teacher's interventions would be based upon guidance as to the appropriate task for the skills of the child to manage rather than the traditional teacher's role of implementing a timed, pre-determined curriculum dictated to the entire group. Discipline would be based on guiding children to resolve problems themselves. Misbehaviour would not be dealt with on the basis of traditional punishments and rewards, rather it was predicated on re-directing the child's attention towards purposeful activity where she had observed success previously.

In this regard the environment had to be prepared prior to the children commencing their time in the nursery. Montessori believed that it was important to attach children's play to activities based upon a child's mindset in the present moment, rather than wilful play through the arbitrary introduction of toys and games simply because that was what was available. Clearly this method imposed its own discipline upon the child and the teacher if applied rigorously. She identified several distinct phases of development, each having its own 'apparatus' scientifically organized by subject, and its degree of difficulty. All materials were displayed on open shelving and were available for free, independent use, or as a sequence controlled by the teacher. It was intended to stimulate their natural instincts and interests for self-directed

A historical overview

learning. Aesthetics are extremely important in a prepared environment because the child chooses the activity, and for that to work, the play equipment and the surroundings must attract the child.

One might contrast Froebel's first kindergarten, located in a beautiful wooded valley in Thuringia, with Montessori's House of Childhood (Casa dei Bambini) in the most squalid district of urban Rome, San Lorenzo. Their differing viewpoints and preoccupations are reflected in this stark disparity. Froebel, in an ideal rural environment, focused on nature's gifts in the development of the child and the outside spaces where they would experience nature at firsthand; an environment which was as close to idyllic as conceivable. Montessori centred her attention on the immediate surroundings, emphasizing its importance in much more practical ways. She did not have beautiful outside spaces, rather there were hard, enclosed streets and alleyways teaming with adult activities of all descriptions. Therefore, the interior suddenly became critical, a sanctuary, just as it had for Owen a century earlier. However, Montessori's system was much more scientific in its make-up, more like a machine for learning in than the proverbial Garden of Eden.

Montessori believed that the first task with very young children was to encourage or train the child to be independent in 'basic tasks', such as toilet training and hygiene, dressing and rudimentary organization. This again required a designed environment to provide for these most basic and practical activities. Beyond that she developed a further set of categories which she called 'sensorial', 'language' and 'cultural subjects' before the child moved on to an elementary curriculum at age six. What were called 'practical life materials' were developed to promote physical coordination, care of self and, most importantly, care of the environment. Materials were created for self-help dressing activities using various devices to practice buttoning, zipping, bow-tying and lacing. Other practical life materials including pouring, scooping and sorting activities, cleaning (washing a table) and food preparation to develop hand–eye coordination were introduced. Slightly more prosaic were exercises such as walking in a straight line, sitting in a chair and climbing stairs, for which Montessori developed a set of small-scaled steps, perhaps creating such child-centred apparatus for the first time. She went on to introduce child-scaled furniture such as play kitchens, devices which are commonplace today.

Following her initial experiments with young children, she extended her research by introducing new materials and further studying and recording the effects of her approach with children of different ages. She lectured all over the world in her later years, publishing her findings and developing further her progressive theories of child development. Today, many hundreds of Montessori schools based on her broad principles exist throughout the world. Although there is no clearly defined Montessori curriculum, her principles have evolved to mean nursery care and education of the highest quality within purpose-made aesthetically pleasing environments which closely match the patterns of children's play as established by this important visionary educator.

In America, the kindergarten movement was spread by the migration of European educational pioneers during the early part of the twentieth century. They commissioned some important examples of modernist architecture, producing buildings which would reflect the growing importance of children in society and the radical humanism at the heart of the kindergarten ideal. For example, the

A historical overview

Montessori's first Casa dei Bambini contained within this Rome tenement block, 1926. This was superseded by more appropriate purpose-designed buildings such as this Montessori nursery at the Karl Marx Hof, Vienna by architect Karl Ehn (1932). Here the building and related play area are integral to the housing complex; indeed, children become the physical and symbolic focus of the community embodied within its form and structure. (Wright, H. Myles and Gardner-Medwin, R., *The Design of Nursery and Elementary Schools*, The Architectural Press, 1938).

A historical overview

nursery buildings at Oak Lane Country Day School, Pennsylvania, by Howe and Lescaze (1929), were designed in a fashionable stripped-down white style of architecture, with flat roofs and parapet walls. The distinctive corner windows gave a soft mixture of north and west light. There was a rooftop playground proposed, an idea which would later be copied in more urban locations.

Kindergarten buildings in the United States further developed this modernistic architectural language during the period up to the Second World War. Walter Gropius designed a fully glazed 'folded' façade in his project for the Caryl Peabody Nursery School (1937) which promoted key kindergarten environmental ideas with opening windows/doors and external canopies for shelter to provide children with easy inside–outside play and lots of sunshine.

During this time kindergarten architecture emerged as an important element of many new social housing projects in Europe. For example, in 1927 Karl Ehn was commissioned to build a new Montessori kindergarten at the vast new social housing at the Karl Marx Hof, Vienna. One of the main courtyards that was enclosed by residential blocks contained a charming Montessori nursery; rather than adults being the physical focus, it was to be young children. Today it is more often the motor car which assumes this central role. The nursery's position right at the heart of the new residential blocks was a symbolic reference to the central role children would have in the new society. Le Corbusier proposed a heroic rooftop play space as part of his Unité high-rise housing block in Marseilles (conceived in the 1930s, built 1947–1952). It included a cooling play pool which, having been recently restored, was in full use when we visited last August.

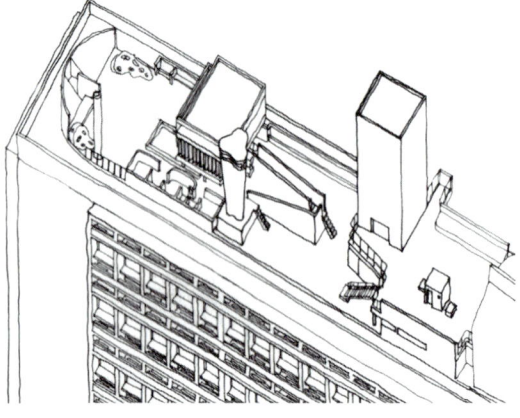

The Marseille Unité by Le Corbusier (1947–1952). The rooftop children's play area and health centre which have been recently restored hold the key to the architect's original view that any housing community should take the needs of children very seriously.

A historical overview

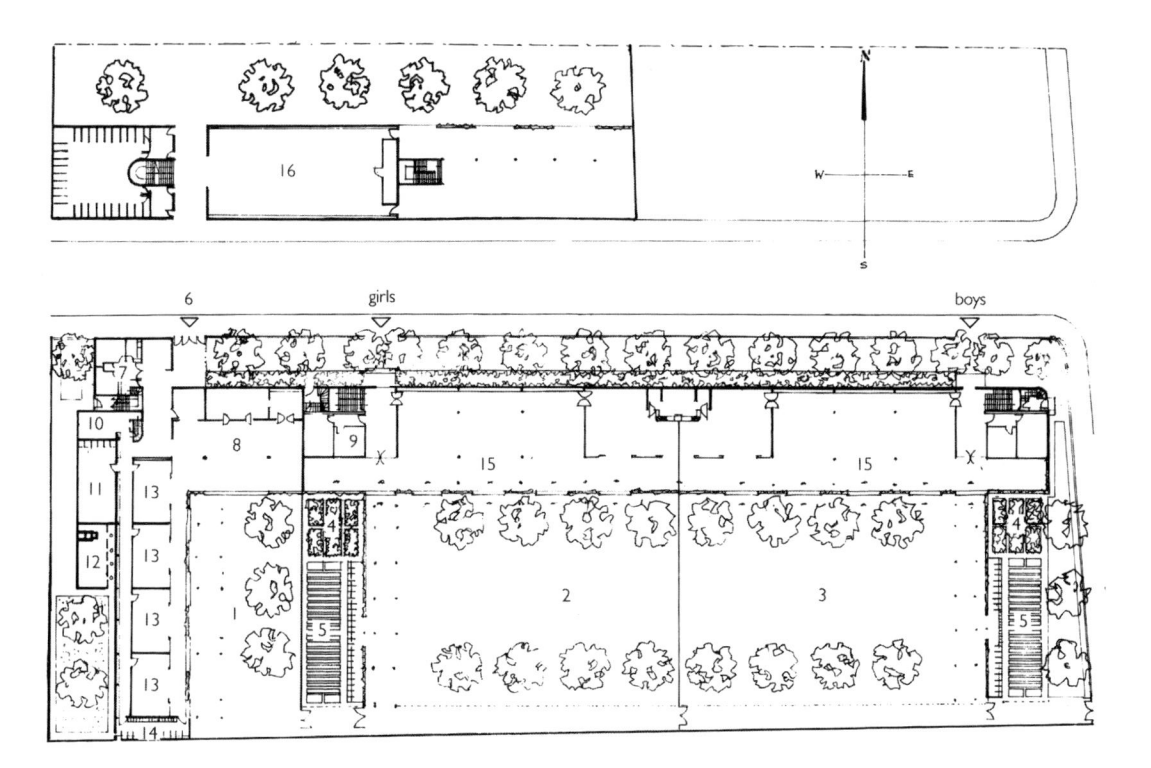

Ground- and upper-floor plan of Andre Lurcat's school at Villejuif. The kindergarten part is contained around its own courtyard, which is separated from the main school by the head's office. It is considered to be one of the first examples of 'free planning' along open-air lines. Key: 1, kindergarten; 2, girls' playground; 3, boys' playground; 4, flower garden; 5, vegetable garden; 6, main entrance to kindergarten; 7, porter; 8, dining; 9, head teacher's offices; 10, medical room; 11, dormitory; 12, plant room; 13, classrooms; 14, WC; 15, junior school with classrooms over; 16, sport/activity space.

Nursery school projects in the Paris suburbs during this period, particularly those by Beaudouin and Lods at Suresnes, helped to establish a French tradition of good, local, state-funded kindergartens. In France in the 1930s, a number of infant schools with attached écoles maternelles were built. One at Villejuif, designed by Andre Lurcat, was a notable example of this new architectural typology. It had a spacious open form with an L-shaped layout. There were four activity areas, which opened onto a secure garden. A dormitory occupied the short arm of the L. The garden became the focus of the new learning curriculum, where the cultivation of fruit and vegetables was very much part of the children's everyday activities.

During the period leading up to the Second World War there were important developments in the United Kingdom. In 1933 the influential Hadow Report on infant and nursery schools addressed the needs of children under seven. Influenced by Montessori and other educationalists such as John Dewey, who raised awareness of the practical dimension of education, the report's emphasis was on activity-based play as the key to social training. It underlined the need for more space than the ten square feet (per child) which was recommended in the design of school buildings at that time. The spirit of the report is best summed up by the comment that 'the ideal infant school is not a classroom but a playground', emphasizing the importance of open-air education.

A historical overview

Écoles Maternelle Ville De Cachan designed by Chollet and Mathan in 1933 established a template for the new French state kindergartens. They were big multi-storey institutions with a solid civic presence incorporating signage and graphics almost on the level of an institutional brand similar to Frank Pick's work on the London Underground. It spoke optimistically of the future citizen expressed in an architecture forged in the spirit of the Modern Movement.

As the century progressed, women were increasingly needed in factories, and by the 1930s there was a concern to produce a cheap, standardized nursery school to satisfy demand, especially as Britain re-armed and many millions of women found themselves working in factories or on the land. One of these projects was commissioned by the Nursery Schools Association and designed by Erno Goldfinger, who went on to build many iconic residential and commercial buildings in the High Modernist style. The interior showed a rather scaleless environment, suggesting that Goldfinger understood very little of the need for child-friendly design. Today, there are many examples of economical pre-fabricated and pre-engineered nurseries, some of a high quality, but many which are inappropriate, failing environments even on the very day they are opened. One wonders how Local Education Authorities

A historical overview

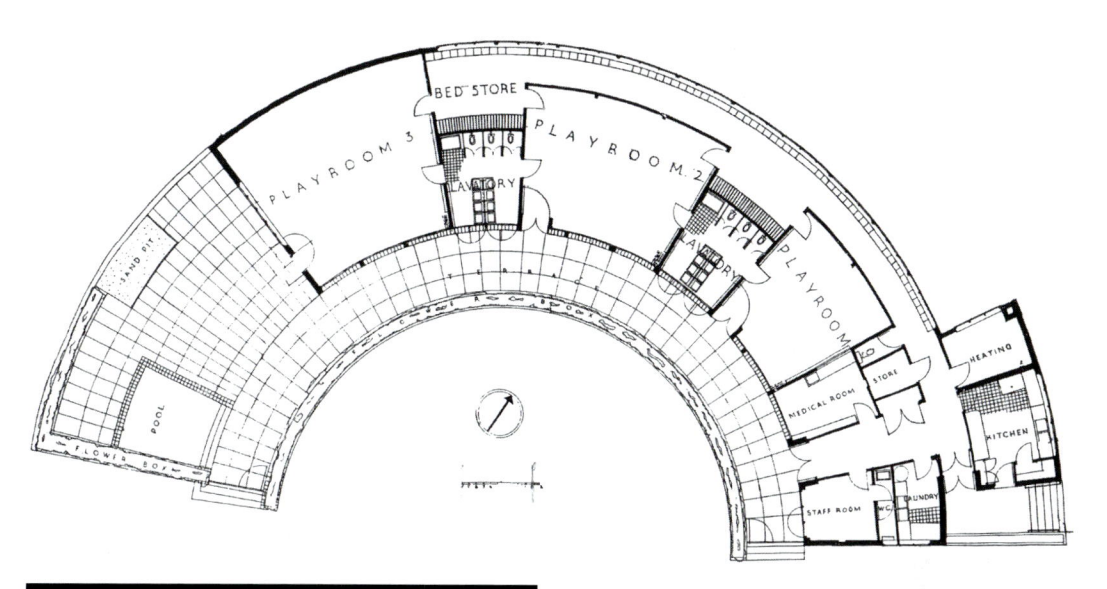

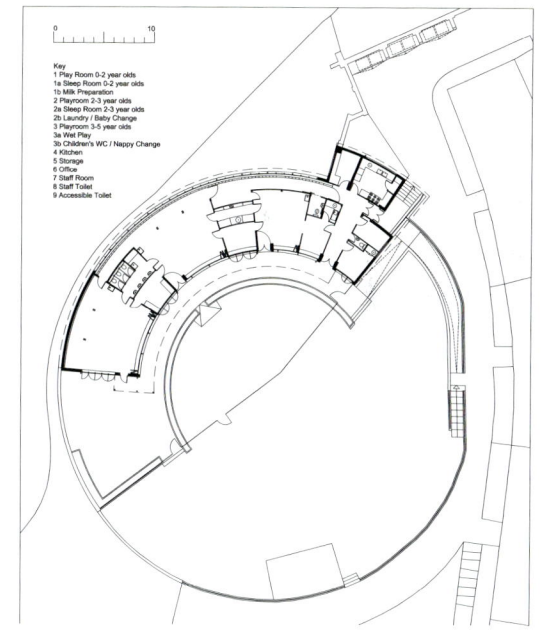

This semi-circular nursery at Kensal Rise, London, designed in 1936 under the leadership of Maxwell Fry adopts an intimate relationship with the semi-high-rise housing in the background. One could imagine parents waving to their offspring from the balcony, keeping a watchful eye out as the day at nursery progresses. Compare the original plan, with the recently refurbished version: internal accommodation has been extended into what was the covered outdoor play area to create an additional playroom for 3–5-year-olds, reflecting the additional role the building has; to provide for twice the original number of children, albeit on a part-time basis. The refurbished playroom shows the sound principles of Fry's initial design with the high-level clestory windows to provide modulated light and ventilation (photograph and lower plan permission of Cotrell and Vermeulen upper plan permission of RIBA Drawings Collection).

A historical overview

can make the same mistakes made by previous generations on the basis of economic expediency.

In 1937 architects Maxwell Fry and Elizabeth Denby designed an elegant semi-circular nursery school. It established an important theme – that of the nursery being close to the home – as it was integral to an experimental social housing project at Kensal Rise in North London (Kensal Rise refurbished scheme). Across Europe, the nursery often became the focus for larger social housing projects during the 1930s and 1940s; Le Corbusier's Marseille Unité was built between 1947 and 1952. This high-rise block was equipped with a roof-top kindergarten complete with paddling pool, integrated climbing ramps and water play features, which formed an exciting futuristic play landscape. Earlier, in 1934, Swiss architect Hans Leuzinger produced an archetype for the modernist nursery school when he designed his, near Zurich. The interior shows modern ergonomically designed children's furniture in harmony with the timber structure of the building itself. It is light, airy and a perfect foil for the rugged mountain setting behind.

In the years following the Second World War, partly for cultural and partly for social reasons, architecture for primary schools took priority at the expense of purpose-made pre-school environments. There was a widespread view that children should be at home with their mothers, while in many other European countries, mothers went out to work. For example, the advanced quality of early-years provision in Italian regions such as Reggio Emilia was mainly due to the mobilization of women as the primary workforce providers, due to the imprisonment of so many Italian men in the immediate aftermath of the war. This necessity, combined perhaps with the balance of power being in the hands of women rather than men, created a culture of emancipation, which was generations ahead of the male-dominated culture of 1950s Britain.

A refurbished interior of the Kensal Rise Flats and integrated nursery, originally designed under the leadership of Maxwell Fry and sensitively restored by architects Cotrell and Vermeulen (2007). The activity space has a spatially generous quality with a variety of ventilating windows which keep it cool in summer. Note the highly practical storage running the entire length of the back wall, with integrated child-height seating.

Even during the so-called swinging sixties, very few nurseries were funded, and Mrs Thatcher's reactionary view, that women (except for herself) should be at home looking after young children kept all but the most vulnerable and at-risk children in part-time, poorly funded early-years facilities, if at all. It is ironic that Britain, which was arguably the birthplace of a coherent architecture for young children, had slipped so far behind the rest of Europe during these critical post-war years, placing the economy and the well-being of families at a significant disadvantage to other European competitors. A case of short-term gain for the few at the expense of long-term gain for the many. It is perhaps not surprising that out of 28 countries surveyed for an influential OECD report in 2002[18], Britain was the worst place in the industrialized world for child happiness and welfare.

Throughout the twentieth century, young architects used the kindergarten as a stepping-stone to greater things. This can be seen particularly during the 1980s, when many German cities upgraded their daycare facilities to gain economic advantage over other competing cities. A particularly expressionistic example is the Luginsland Kindergarten, the so-called 'sinking boat', in Stuttgart, designed by Behnisch and Partner as shown on p. 29. The City of Frankfurt commissioned over 20 state-of-the-art buildings for young children, including the Frankfurt Greisheim-Sud Kindertagesstatte, which provides full daycare for over 200 children, together with extensive after-school facilities for older children. This initiative gave young up-and-coming architects such as Toyo Ito and Bolles Wilson the opportunity to develop these low-budget buildings with verve and ingenuity prior to going on to highly successful careers working on large, mainstream architectural projects. There would appear to be an important correlation between childhood perceptions and the ability to make architecture which is playful, innovative and centred on the users' sensory relationship to space.

The most widely quoted architect said to have been influenced by his childhood experiences was Frank Lloyd Wright. On many occasions Wright described the significance of the Froebel 'gifts', the first set of children's building blocks which were developed for Froebel's school at Yverdun in the 1830s, and used by Wright as a child. There have been many studies which relate the form and proportion of Wright's architecture to these construction toys. For example, Wright's designs for the Avery Coonley Playhouse illustrate a planar architectural style which emphasizes horizontal layered form making, which in a certain light can seem like it is formed out of Froebel building blocks. The building included a kitchen with child-height worktops and a stage with a dressing room. The fireplace behind the stage acted as a permanent backdrop and the symbolic heart of the building.

From a historical perspective, the three large stained-glass windows at Avery Coonley, which Wright designed in colourful abstract form, were important exemplars. Rather than using obvious child-like imagery, such as Mickey Mouse or Rupert Bear shapes, Wright instead abstracted the design into playful cubistic shapes which could be interpreted in a number of different ways; are they balloons, or is it an American flag fluttering in the wind? That children could use their own perceptions to imagine and play in a more creative way through abstract architectural design was, and remains, an important kindergarten concept.

A historical overview

Taking this logic literally back to the Froebel gifts, Wright himself ascribed the triptych stained-glass windows to the seventh Froebel gift. He claimed that these circles and squares of brilliant primaries 'interfere less with the function of the window and add a higher architectural note to the effect of light itself'. They form what Wright called a 'kinder-symphony', once again evoking Froebel's kindergarten education:[19]

> At the Centennial in Philadelphia, after a sightseeing day, mother made a discovery. She was eager about it now. Could hardly wait to go to Boston as soon as she got home – to Milton Bradley's. The Kindergarten! She had seen the 'Gifts' in the Exposition Building. The strips of colored paper, glazed and 'matt', remarkable soft colors. Now came the geometric by-play of these calming checkered color combinations! The structural figures to be made with peas and small sticks: slender constructions, the joinings accented by the little green-pea globes. The smooth shapely maple blocks with which to build, the sense of which never afterwards leaves the fingers: form becoming feeling. The box had a mast set upon it, on which to hang the maple cubes and spheres and triangles, revolving them to discover subordinate forms.[20]

The ideas that Wright pioneered in his early Prairie House phase are inextricably connected to his play as a child with the Froebel blocks. Elsewhere he expressed his childhood excitement at the work with the blocks, when his mother had finished her housework and he could sit down with his brother at a low mahogany table with its polished top, and the three of them would 'work' with the 'gifts'. Here one senses that subtle balance between what could be described as play and work. Not only is the child held back from time spent with the blocks (thus creating a dynamic tension or anticipation), which enhances their magical qualities, but the sheer act of hand and eye coordination utilized in their very spirit helps this growing coordination to develop, and in the hands of the young Wright, thus in relation to his architectural genius, form becomes feeling.

During the period of the immediate post-war years, kindergarten architecture took a back seat to the re-construction of infrastructure generally. While the United Kingdom and United States did little to provide a coherent early-years architecture until relatively recent times, in Spain and Italy things moved forward from the beginning of the 1980s, where numerous examples of good kindergarten architecture were built in those countries during the decades following.

In Italy, the pioneering work of Loris Malaguzzi in Reggio Emilia during the 1940s and 1950s created the foundations for an entire city-wide system of municipal pre-schools of the highest architectural quality. In Modena, children and families took precedence over the needs of the motor car, which was somewhat ironic as this is the home of Ferrari. Malaguzzi believed strongly in the importance of architecture to young children, which he described as 'the third teacher', after the kindergarten carer and the parent. There, designs such as the San Felice Nursery and Preschool by architects ZPZ Partners provided a template of excellence for many architects to

A historical overview

follow in the use of colour, space and materiality. The book on Reggio entitled *Children, Spaces, Places, Relations*, first published in 1998, remains relevant in its theoretical and poetic approach to children's spaces. It emphasizes the relationships which emerge as a result of the time children spend together in nursery.[21]

Spain's Educational Reform Law of 1990 was established in response to the existence of strong social pressure to create nursery and infant schools of the highest quality. Since then, Spain's system has developed a culture of democracy, autonomy and egalitarianism which is reflected in kindergarten examples such as the Bilbao Sondika Kindergarten by Eduardo Arroyo and No.mad arcquitectos, where up-to-the-minute materials are used which create a rich, futuristic environment for play, full of modulated light which absorbs the reflected light from the mountainous topography beyond.

Professor Helen Penn makes the point that the Spanish and Italian nurseries are places that foster a children's culture, providing sufficient high-quality space and opportunities for children to explore and produce their own means of expression.[22] She cites the nurseries in Barcelona as playing an important role in sustaining the local Catalan culture through, for example, language (the children speak Catalan rather than Castilian in the nurseries), and food, which is given high priority in both the Spanish and the Italian nurseries.

Most importantly, commissioning bodies in Spain and Italy recognized that kindergartens can be places which introduce children to a rich aesthetic culture, painting, sculpture, music and most importantly architectural space, which they believe are fundamental to healthy development. With some exceptions, in the United Kingdom all too often the nursery is viewed merely as a business or a service delivery 'system' for incarcerating young children. As an architect myself getting towards the end of the painful process of completing three children's centres in London, I am only too aware of the lack of recognition of the importance of good design, and the often-bewildering level of ignorance in relation to these longstanding values.

This brief summary of the nursery/kindergarten building type in history would be incomplete without mention of the often contrary educational pioneer who developed a particularly architecture-focused approach to early-years education during the twentieth century, Rudolph Steiner.

Steiner, like Froebel before him, believed that pre-school children needed to play rather than engage in formal educational tasks in order to awaken their lives. However, he added that this awakening needed to happen in harmony with the natural world. His ideas instigated a powerful synthesis between architecture and pre-school education. His key concept was that one inorganic form would be added in sequence to another to create a system which resembled an image of growth, incomplete and therefore dynamic, and natural. Most importantly, an architecture which eschewed the conventional geometries of the right-angled building form. It was this sense of spiritual metamorphosis, an embodiment of the process through which the pre-school child passed, which made the Steiner approach so evocative.

The often organic retro design used in many new Steiner nurseries can be seen as a counterpoint to the recent tendency towards the adoption of modernistic design in high-end kindergartens. With its turf-covered roof and

A historical overview

A contemporary Reggio nursery located in Modena. Scuola dell' Infanzia San Felice represents the best pre-school system anywhere, with its sophisticated arts-based curriculum, high-quality teachers and an architecture which is not only practical, but is also rather beautiful. Based on the idea that children learn most from social interaction with their peers, the building is set-up to promote cultural and social interaction at every opportunity. Each centre has accommodation for 75 full-daycare children aged 3–5 in self-contained home base units of 25 children, each unit containing its own art room and music room as well as the usual play areas. These are large buildings compared to their UK equivalents; each child has approximately 8 m² of play/activity area. There is in addition a baby room for 15. Plan key: 1, activity areas – toddlers; 2, babies' sleep and activity area; 3, open court; 4, main entrance; 5, all-ages dining area; 6, kitchen; 7, staff area; 8, library; 9, administration offices; 10, children's home-bases; 11, atrium.

A historical overview

It is noticeable how the entire school has internal space to meet socially, either informally within the piazza, metaphorically the town square within this city for children, or on a more formal basis, for example within the communal dining areas and atelier spaces. The entire conception is predicated upon Loris Mallaguzzi's original ideas formulated during the 1950s and 1960s in developing a children's culture based on research into children within childcare services. Here, key words and metaphors are broadly understood as the fundamental starting point for any early-years architecture: 'Overall softness; relation; osmosis; multisensorial; epigenesis; community; constructiveness; narration; rich normality'. Within this sensitive understanding I would add that the children are experiencing the most fundamental ordering of their lives through the actual form of space within which they play out their early years. As a consequence, when they reach school, aged six, they are ready to learn as they have a deeply marked memory map of the social structures which will dictate their lives. Architecture by ZPZ. Quote from *Children, Spaces, Relations*, published by Domus Academy Research Centre, Via Savona 97, 20144, Milan.

Shown are the view from the garden – there is a sense of scale and order which is apparent both externally and internally (page 136); the piazza, which is a two-storey volume which enables upper decks to be accommodated for children to rest and relax (page 137); a sleeping area with playful lighting and purpose-made couches – this allows children to have good rest time during their full day in the centre (above left); a typical home base area with integrated storage and upper-level child areas.

thick, highly insulated tapering walls, projects such as the Nant-y-Cwm Steiner School in Llanycefn, Wales, should perhaps be viewed as a refreshing symbol of a more sustainable future in this ominous age of climate change.

A children's world focusing upon a garden environment seamlessly knitting into the internal world of the nursery building (as opposed to the school which was and remains an environment which is largely autonomous from its external grounds), which defined the innocence myth for early years from the eighteenth century also helped to establish the first design principles: namely that the ideal nursery environment would be open to the outdoors, to the extent that its internal spaces would provide free run-in–run-out access to the surrounding garden setting and that it would also reflect the natural world in its internal use of materials and its optimization of natural light and ventilation.

This latter idea, the use of sunshine and full-width opening doors to optimize natural ventilation, was a principle which coincided with High Modernism in architecture (from the 1920s), with its ideological emphasis on the health-giving benefits of sunlight and fresh air. With the new steel technologies that enabled wide-span structural frames and the principles of the so-called free plan, allowing large areas of glazing with big, openable

A historical overview

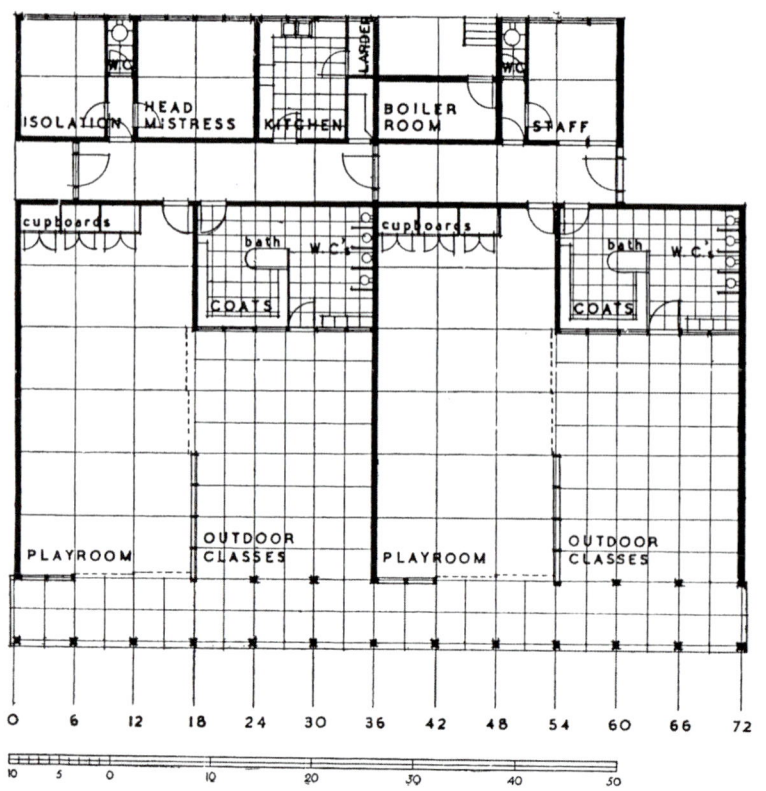

Pencil and crayon internal perspective for the expanding nursery school designed by Erno Goldfinger and Mary Crowley (1937); and the plan drawing illustrating the offset form incorporating alternate classroom and open court arrangement.

A historical overview

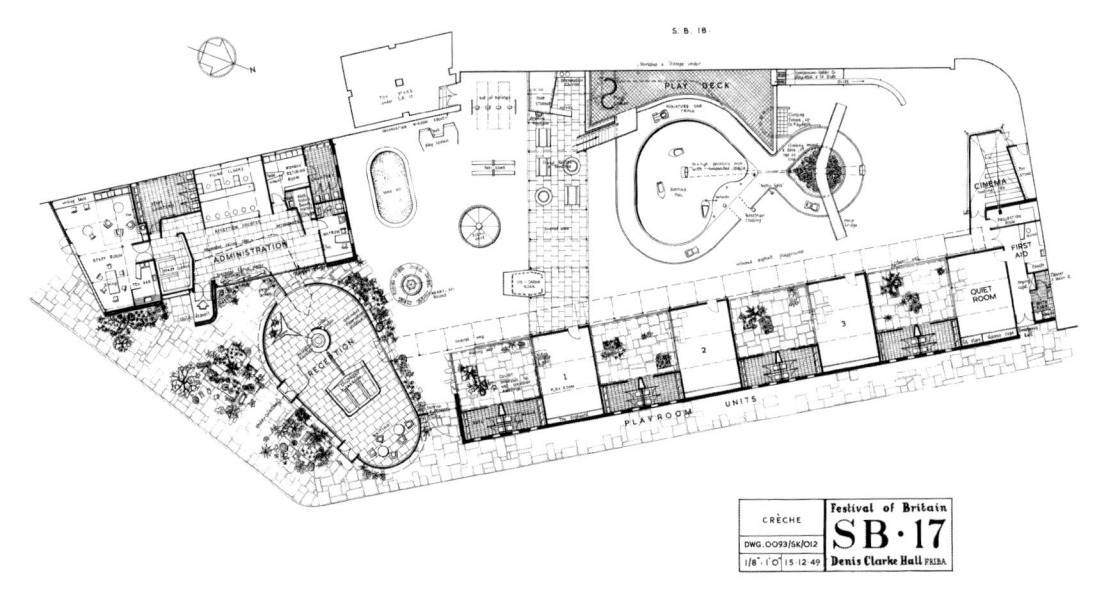

This charming crèche was designed for children visiting the 1950 Festival of Britain on London's South Bank by Denis Clarke Hall. It featured a playground, miniature race track, boating pool and a small cinema.

window-doors, the nursery was an ideal vehicle for many of the pioneering early modernist designers such as Le Corbusier, Mies Van Der Rohe and Erno Goldfinger to cut their teeth on.

The new nurseries would be purpose-designed, warm and dry rather than being located in the damp basements of existing tenement buildings; they would be secure and enclosed yet also low scale and designed to the scale of the child, a further gift of modernism to this new building type with its dissected pavilion structures and low-level flat roofs. However, Modernism was also a somewhat austere aesthetic which promoted clean, hard lines and often brutally honest materiality, such as exposed concrete or structural steel.

A historical overview

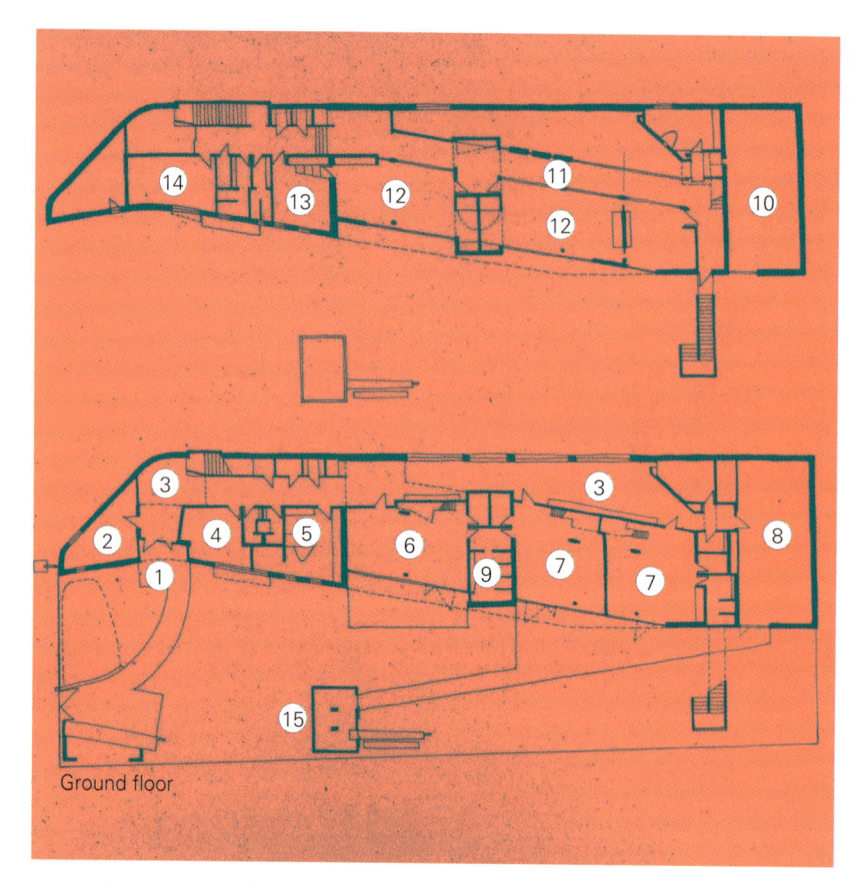

Ground floor.

Ground floor

A historical overview

The Frankfurt Kindertagesstatte, by architects Bolles Wilson. The planning encourages movement. The building itself is designed in such a way that it grows as the children grow; babies and young children up to the age of six are located at the low, narrow end close to the main entrance and the kitchen, while older children and facilities for after school are at the higher, wide end of the building (page 140). This makes the scale of each age-related area more sensitive to the needs of the children using it. This is a building of great appeal to children and of which the users and the surrounding community feels proud. Twenty years on from when it was built, the Kindertaggesstatte is still attractive and free of graffiti – testament to the care it evokes in the locals.

Key to the plan: 1, main entrance; 2, head teacher's office; 3, hall and stairs; 4, office; 5, kitchen; 6, babies' base; 7, children's base; 8, performance hall; 9, children's washroom; 10, void over performance area; 11, balcony over activity space below; 12, after-school club for older children; 13, staff; 14, interview room; 15, garden store.

Shown are: interior of the toddler's area prior to furniture and people – perhaps the architect thought this might spoil the purity of his architecture; the main staircase, which is used by children – the architects have cleverly positioned a red symbol to indicate to pre-literate children that the staircase is potentially hazardous – colour form language; the view from the garden court, with the children's toilet block picked out in blue (see also page 158).

CHECKLIST 4: KEY LESSONS FROM HISTORY

1 Pre-twentieth-century terminology like 'kindergarten' and 'nursery' has somewhat biblical connotations relating children to the idea of the Garden of Eden: it remains a moral concept today, but not necessarily a religious one.

2 Experimental pre-school systems have initiating ideas like 'child-centred learning', 'open teaching' and the 'outdoor classroom': these and other early-years inventions anticipated many later developments in general education.

3 The ideal early-years principle is to bring about a convergence of education and architecture to create transformational environments for learning: designers of these buildings must be fully conversant with relevant educational theories.

4 The two sometimes conflicting early-years principles were (and still are) keeping the child safe and providing challenging physical activities: what follows from succesfully balancing those ideas is accelerated child development.

5 Everyone involved in the development of the new building must have the courage to create spaces that challenge children: the early-years movement is, and always has been, an ongoing development of new progressive environments, often in conflict with orthodoxy.

6 The origins of the kindergarten were based on nature, with an emphasis on sensory learning. Today there is a balance to be struck between traditional values and new forms of learning: while digital play is the new reality, we must not forget the founding principles.

7 The first early-years designers believed that young children understood through a symbolic mystical language which at its most informative utilized metaphor and analogy; this was often derived from natural phenomena and/or mythical stories. It is still valuable today as inspiration.

8 Most early educators emphasized that the relentless physical activity and restlessness of the child should be sustained and directed towards developmental goals, rather than it being crushed by a sense of overbearing discipline and constraints.

9 The primary environmental idea that indoor activity spaces should open directly onto the gardens remains; to provide run-in–run-out play is and always has been a key idea, thus extending the field of learning.

10 The child as a self-motivated empirical scientist was a strong and abiding image; in the second part of the twentieth century mysticism lost traction and brought the science of architecture into focus more clearly as part of the 'laboratory' concept.

11 Although it is a somewhat subjective idea, for many, aesthetics are an important concept in the prepared environment: the child chooses the activity, and for that to work, the play equipment and its surroundings must be attractive.

12 Teaching the child to be independent in 'basic tasks', such as toilet training and hygiene, dressing and rudimentary organization is a core idea; this requires a designed environment to allow these activities to take place. For a child every room is a place for learning.

13 There are many examples of economical prefabricated and pre-engineered nurseries, some of a high quality; unfortunately, most are not and diminish the value of young children in our society, remaining a short-term and inappropriate exigency.

14 Kindergartens and nurseries can be places which introduce young children to a rich aesthetic culture: painting, sculpture, music and, most importantly, correctly designed architectural space. Complex environments promote natural curiosity in the child, and from this, learning develops effortlessly.

Chapter 5

A nursery brief

A machine for learning

—— Adjacency diagrams and rooms list ——————

This brief relates to a specific project we are working on in the town of Jurgow, Poland. There is not – or should not be – a standard brief for every nursery; each is specific to a particular site, relating the prevailing environmental conditions to local needs, and to the specific requirements of those who are going to be using the facility. Nevertheless, it is valuable to bring the experience gained working on other early-years projects, to inform the development process and to promote the highest standards of environmental quality here in the Tatra Mountains.

In practice, what works and what does not work has been tested and evaluated by the authors within the framework of numerous other projects currently in use. Designing for children is an evolving science which is usually constrained by practical issues and requirements, which we set out systematically below. Needs change and evolve over time, and the mark of a good kindergarten is its capacity to develop along with its users with a loose-fit, long-life concept. It should be capable of adaptation, change and growth over time. Most early-years buildings are usually a grand compromise and this should be borne in mind when reading this generic brief. Often, decisions will be based on the careful ordering of competing priorities; sometimes it is simply what we can afford. However, it should not be an excuse for poor decision-making at the development stage.

Here, one also has to take into account the cultural background of any new development, what parents have been used to, and what they are prepared to 'buy' in terms of the hopes and aspirations they have for their children. For example, we recently visited a centre in Viborg, Denmark. On entering the children's garden we were quite surprised to see 4–5-year-old children playing with a bonfire, throwing leaves and twigs on top, in a totally relaxed way. I asked the centre manager about this: wasn't it dangerous for such young children to be left tending a fire? She responded that because they had a long 30-year tradition in Viborg of daycare for early years, parents, even grandparents had attended and engaged with similar activities, so the children saw it as normal and behaved responsibly. This, she believes, is part

A nursery brief

Kilburn Grange Park Adventure Playground, the visitor centre with its 'folded' roof.

Activity centre interior, which acts as an after-school club.

A nursery brief

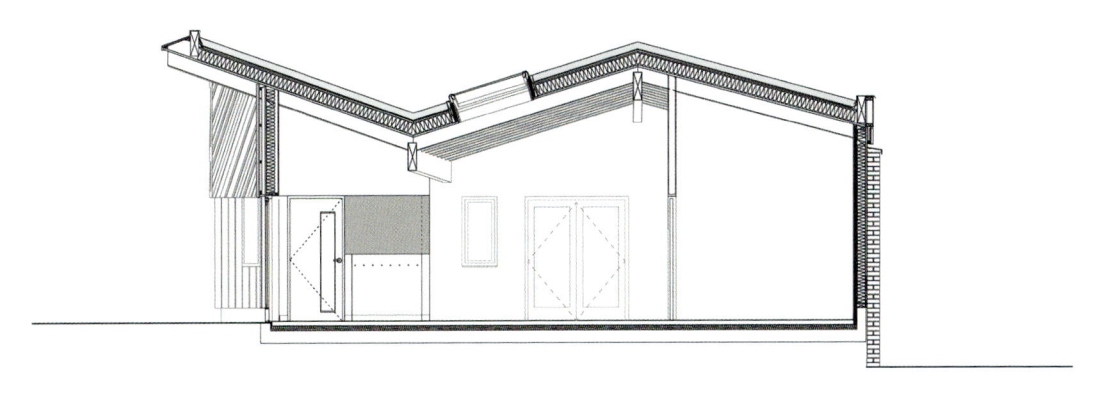

Section 2

Kilburn Grange Park Adventure Playground
erectarchitecture.

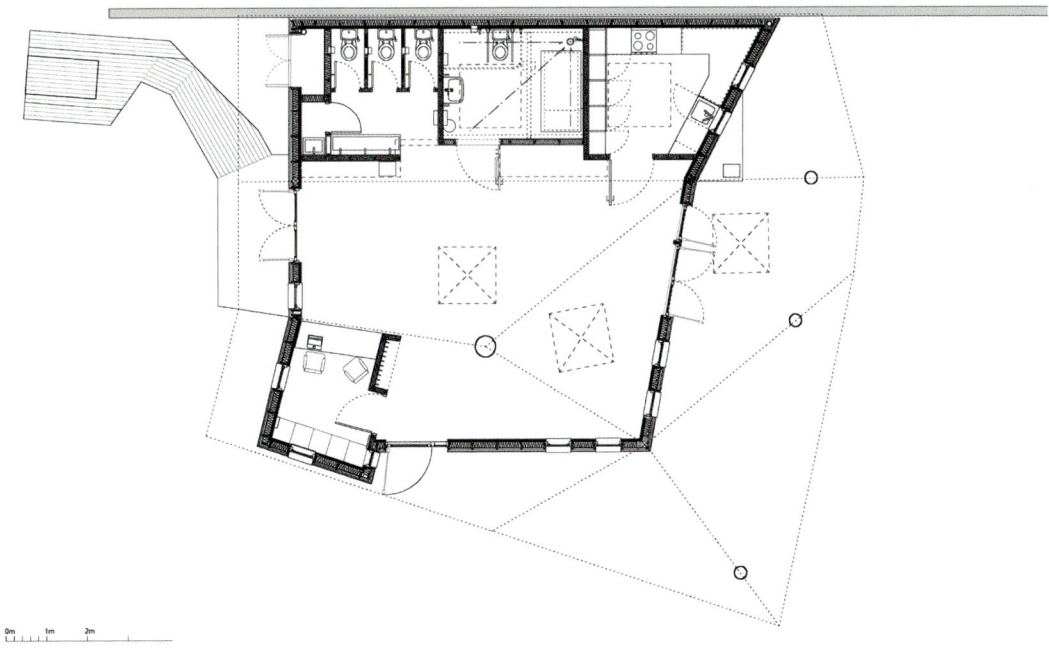

GF Plan

Kilburn Grange Park Adventure Playground
erectarchitecture.

Cross-section and plan.

A nursery brief

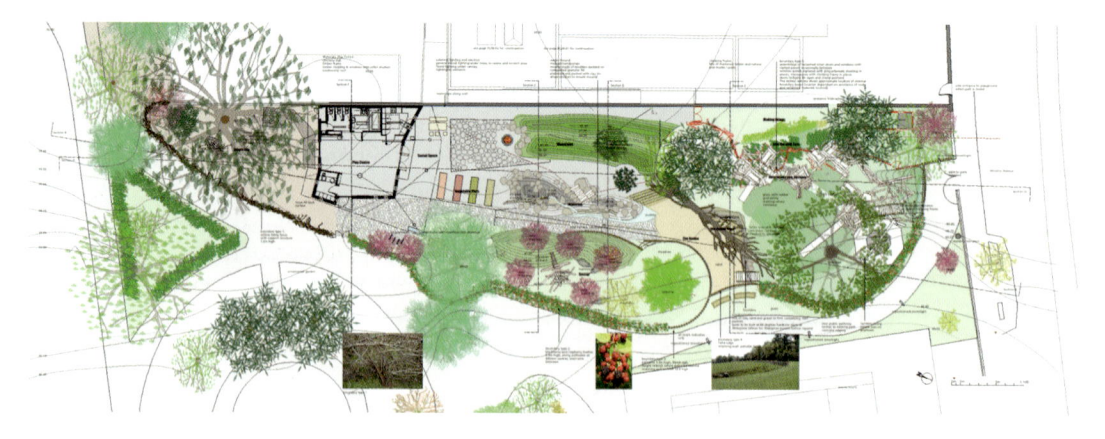

The site was a disused depot, itself built on the remainder of a Victorian arboretum much of which has been retained. The scheme celebrates playing in and around trees.

This exciting play feature comprises a series of platforms rising up into new and existing trees.

The water feature provides some of the best sensory play for children, but is only used on warm, sunny days.

of the culture of Viborg families; in other towns and cities without this heritage it would take time to subsume an early-years culture into normal family life.

This brief deals almost exclusively with what might be described as the 'nuts and bolts' of the new childcare building. Subjects such as the ideal centre size, the precise contents of rooms such as toilets and kitchens, are all described systematically. Details such as the position of toilet soap dispensers and the height of the kitchen servery are presented, to provide a comprehensive list of items to be considered. You also need to match the list to what is available locally. For example, is it even possible for you to get hold of a macerator for the disposal of soiled nappies in Jurgow right now? It is a starting point for further discussion, where high aspirations should be tempered with financial stringency. However, the proposals must be fully understood. Often the problem with such detailed and precise guidance is that it suggests

---

Kilburn Grange Park Adventure Playground, London, designed by Erect Architecture. The inspiration for this new children's play area and activity room is the original adventure playground movement.

Adventure playgrounds originally developed in Denmark more than 60 years ago as a conscious move away from the hard, sterile environments provided in municipal parks at that time. The aim was to create a softer, more natural woodland environment for urban children. Health and safety concerns largely killed them off in the United Kingdom during the 1980s, to be replaced by more predictable 'off-the-peg' installations which are the mainstay of most children's play areas today. Adventure playgrounds were messy: they had a malleable quality which enabled children to move around elements such as logs, water and sand to modify the landscape and create their own play zones. They utilized existing trees and landscape features for climbing and swinging.

From the outset, it was important for the client and the architect, Susanne Tutsch of Erect Architecture, that the scheme was well embedded in the local community; children were involved in the design process, in part to generate a sense of ownership. Intensive workshops both in the school and on-site developed key ideas about the nature of play.

This interaction helped the architect to understand the desire children have both for risk and exploration and for secret or hidden spaces. In response, the architects developed a sequence of distinctive yet fluid spaces of varying sensory quality across the entire site. Crucially, the area located furthest away from the building – the construction zone – is a child-only area, used for making things and enclosed by a full-height wall of old doors. The doors give it a slightly odd anarchic feel, like a Louise Bourgeous art installation lost in North London, or a post-war bomb site.

Overall the design is developed as a conventional sequence of 'outdoor rooms', with varying degrees of enclosure. The activity building is the lynch-pin around which everything else flows. Its distinctive tree-like roof stretches out into the outside play areas, intimating safety and enclosure, while also encouraging less-confident children to go out and explore. Children are allowed to lose themselves across a diverse range of spaces. Although Tutsch describes it as 'controlled chaos', it is clear that supervision by full-time play workers is enabled by a subtle axial arrangement of the outdoor rooms in relation to sight-lines from the main building; only the construction zone is hidden.

The menu of areas reflects both the child consultation sessions and the architect's obvious affinity with the adventure playground movement. The schedule includes vegetable plots, a meadow orchard with fruit trees and the 'mountains', an area of steeply banked escarpments adjacent to the bonfire area. There is a water area constructed of large rocks with mud and gravel sandwiched between; water flows down from the highest point to create an eroding mound in which children can create channels and pools.

The most exciting elements are the treehouse structures that literally hang off newly planted and existing trees. There is a fallen tree spanning the 'ravine', a two-metre cut in the ground. High above is the galleon, a series of walkways and ladders connected to the ground with a wobbly bridge. The architects worked with play equipment specialist Apes at Play, which fully engaged in the witty and playful spirit of the scheme; for example, they found an old piano which is now fixed securely eight metres above the ravine on the galleon's highest viewing platform.

A nursery brief

there are no local decisions to be made. This is not the case, and these recommendations must be considered carefully to ensure that they suit you, the users.

This brief should not just be concerned with functionality. One must also consider important child-oriented spatial concepts. Things are in reality never straightforward when designing for children. For example, it is now widely accepted that the level of spatial complexity within the building aids the development of the young child. It works like this: a complex scene invites exploration more than one which is simple and immediately readable. Just as we experience reward when doing a brain-stretching crossword, a play room with corners and hidden places will hold more fascination for the young child than a straightforward one. A complex scene invites exploration and fascination more than one which is simple and easily readable. From this state of fascination the child learns. In their studies, researchers have proved that by varying a room's features, aspects such as the ceiling height, its slope, fenestration and colour significantly aid the fascination and engagement of young children.

Solid research indicates that if it is possible to engage several of the senses, the greater will be the positive effect for children using it: the use of multisensory materials with a variety of colours, textures, lighting effects, sounds and aromas make the building a true learning environment, because the children's development is significantly advanced as a result of this richness. In child development circles it is often referred to as 'affordances', because it describes our ability to see what an object or a surface or a fixed or moveable feature within the environment can offer the child in play terms. Children learn through play. That cardboard box isn't just a discarded container, it can be successively part of a den, a fire engine, a spaceship, a climbing obstacle, etc.

The pattern of young children's play, as opposed to older school children's, is often characterized by free flow from activity to activity, transforming as it goes. Robin Moore, accompanying young people through 'childhood's domain' in the city is struck by the level of discrimination and

Cherry Lane Children's Centre, West London, designed by Mark Dudek Architects (view of baby unit, inside and outside); the varied plan form and related intricate roof shapes aid child development because a complex scene encourages exploration; this is analogous to an adult doing a brain-expanding crossword or exploring a complex medieval Italian hill town.

A nursery brief

Amy and Grace engage in cooperative play from an early age, the younger child constantly checking her progress against that of the older; this socialization process is a fundamental benefit gained by attending a nursery where play between children is facilitated by the particular spatial qualities of the early-years environment. This should include places where children can 'withdraw' from the hubbub of the main activity spaces if they wish.

inventiveness they show in using resources.[1] Watching children at play in a kindergarten, he observes, one sees this same psychic complexity at work. Younger children tend to play independently of each other while in the presence of others, moving from project to project. Cooperative play emerges later, but can be as early as 16 months.

There are clear implications for the design and running of the kindergarten, to foster and afford this fluidity and ingenuity of play and exploration and to realize that for much of the time, young children will not necessarily spend their time in long periods of focused activity. This is normal, expected and positive. Therefore, play items and resources should be readily available. They should be changed and amended frequently. Even the most inventive interactive water feature will lose its appeal after the fourth or fifth experience if it is fixed and incapable of change. It is also a great advantage if the building itself has an in-built flexibility. We imagine that the three home base/activity areas at Jurgow might be dedicated and designed for different activities. For example, one might be for wet, messy play; another for more contained, clean activities such as reading, computer work and dancing; while the third might be used for quiet floor activities, with a sleeping facility for younger children. They can move through these themed areas in controlled groups at different times of the day.

Sometimes quiet separation is desirable, hence the enclosure of space to form dedicated rooms (see the schedule of accommodation later in this chapter). However, spatial fluidity is also enhanced wherever it is practical, hence the large openable doors between the home base/activity areas and the mini piazza, and similarly between the training room and the parent–infant rooms on the first floor. Visual transparency is also an important concept as much for child protection issues as for the overall semi-open ambiance required architecturally. The client specifically asked for views of the adjacent public park through the activity areas on entering the mini piazza. So,

Sometimes quiet separation is desirable; however, the requirement for staff and parents to keep a watchful eye on children in care is important, and these low walls around the sand-pit enable a sense of separation while maintaining visual contact for the supervising adults.

for example, large openable door/window panels with glazed clerestory panels above will aid the desired lightness and transparency. Similarly, we indicate vision panels between activity areas and children's toilets, which is common practice in most European centres.

However, there must be order amidst the spontaneity of play desires. Child-accessible storage is a must, so that children can take things such as paper and paints and, most importantly, learn to put them away after use. The so-called HighScope curriculum, which we understand will be adopted here, has well-established requirements which will be specified in storage terms.[2] There will also be storage for non-child accessible resources and for larger play items which may not be in use all of the time. Often, a toy library will be included. Hence we will not refer to the early-years home bases as 'classrooms'; rather, we will use the term 'activity area', which reflects this fluidity as opposed to the more static nature of older years school 'classrooms'.

Jurgow is high up in the mountains, at least 2,000 metres above sea level. It is very cold during the winter, with high levels of snowfall. Therefore the roof profiles for the new building are to a certain extent dictated by the weight of snow, and should be not less than 25° in pitch. Because of the extreme winter cold, the children tend to stay indoors for much of the time, although technologically advanced winter ski clothing as seen previously in many Scandinavian projects may, if provided, help to modify this behaviour to some degree. Nevertheless, the need to provide as much internalized space as possible was a primary requirement in the client brief. Therefore the overall form of the building is compact and highly insulated, with a large internalized children's playground, the so-called 'mini piazza'.

Providing the right environment for a child is vitally important, since young children require support of a very particular type in order to develop as happy, confident individuals. What may appear as a gentle flight of steps to adults can seem like a rocky mountain face to a small child – and while this

A nursery brief

Amy safely negotiating a stair aged 1.2 years. The child learns that the world is naturally uneven and multi-dimensional. Her enthusiastic aspiration to explore new and challenging areas should wherever possible be supported by the adult facilitating access in a safe way, yet in a manner that feels independent to the child.

may be a welcome challenge to bored three-year-olds, if the feature is unsafe it will simply alarm their parents and carers.

On the other hand, an early-years environment with few level changes and no stairs is a dull, anonymous and even unnatural landscape. Stairs are, after all, frequently experienced by young children in the home and in other buildings such as libraries or shops. Climbing on and around a staircase can be a pleasurable experience for young children and, if designed safely, will enable the building to be planned over more than one storey, minimizing the built-site space. In addition, stairs and other architectural features can really help to develop the child's physical awareness and their mastery of the real world and all of its unevenness.

For financial reasons there is no lift/elevator in this two-storey building. Most children's centres within the European Community would not by

A nursery brief

The curvy bench at Stanley Infant and Nursery School is positioned by the cloakroom wall, transforming the corridor space into part of the child's activity pathway (designed by Mark Dudek Architects).

law be allowed to function without a lift. However, it is simply not possible in this situation to provide one and consequently the design of stairways takes on more significance. They should be gentle, broad and easy to use, with both child- and adult-height handrails fitted. They should be well lit and colour-coded to provide child-oriented legibility and easy wayfinding. Thus the stairway should become part of the 'promenade' and add to the general enjoyment of the building in use.

Similarly, it is important to provide the environment with other child-friendly features which help them in wayfinding, which is important in environmental awareness development; put simply, being able to remember their way around the building. Note the playfully curvy bench in this nursery. Some centres adopt symbols to denote three different types of child- or non-child-accessible rooms: the doors might be painted green to denote full child access (or a circular window is sometimes used); rooms which are accessible with adult supervision would be painted blue (or have a triangular window) and rooms such as the kitchen which is non-child-accessible at all times would be painted red (or have a square window). Aiding wayfinding and legibility for children (and adults too) through child-centred design is an important requirement. There are many ways in which this legibility can be promoted through architecture.

As with any public building, and particularly one for young children, security is an important obligation for those who run it. Therefore the main entrance at Jurgow Nursery is secured by locks which can be remotely released by a supervising person inside the building. It is anticipated that the administrative office, which is adjacent to the main entrance, will be constantly manned and will maintain visual control over these entrance doors. There are views from the office to the entrance doors and there is a reception desk outside the office immediately adjacent to the main entrance in the mini piazza. Parking will be located in the square close to the main entrance, which is positioned on the west side of the building. The street which runs parallel

Mark Dudek Associates
education design

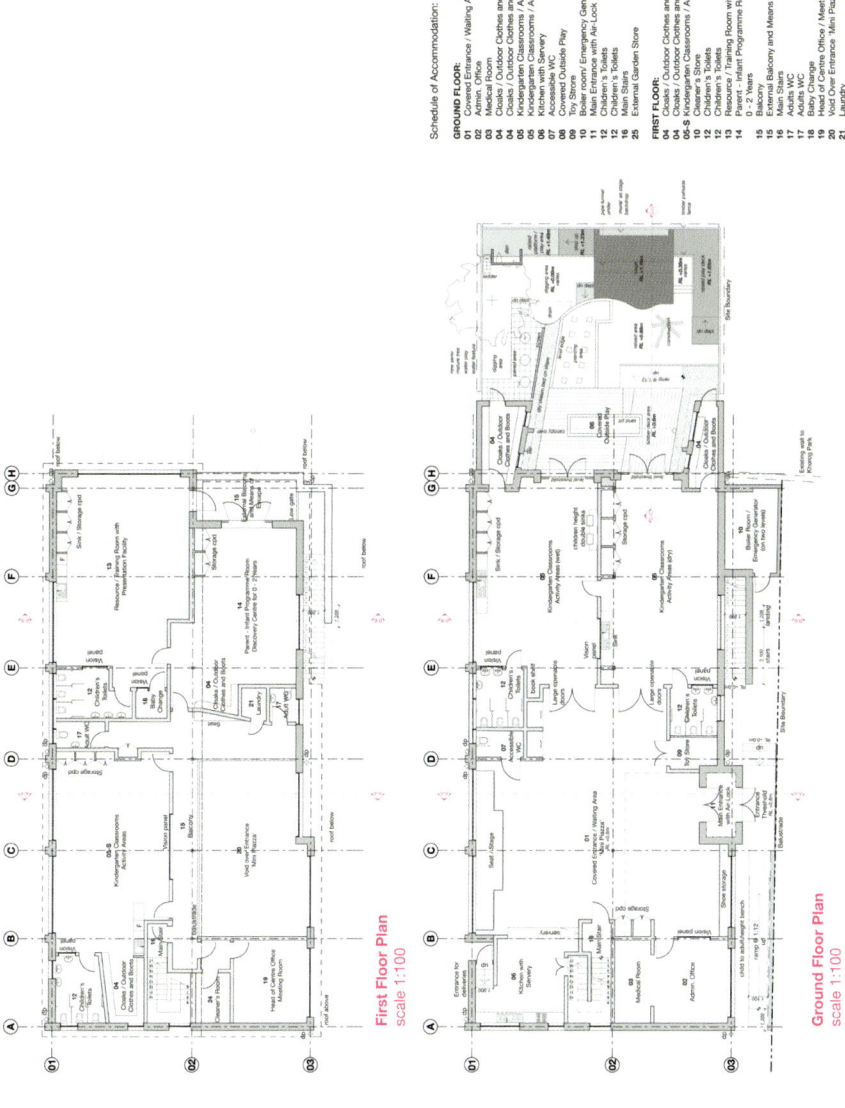

Schedule of Accommodation:

GROUND FLOOR:
- 01 Covered Entrance / Waiting Area - 'Mini Piazza' — 155.9m²
- 02 Admin. Office — 18.9m²
- 03 Medical Room — 6.3m²
- 04 Cloaks / Outdoor Clothes and Boots — 9.1m²
- 05 Kindergarten Classrooms / Activity Area — 71.1m²
- 06 Kitchen with Servery — 72.7m²
- 07 Accessible WC — 25.1m²
- 08 Covered Outside Play — 3.8m²
- 09 Toy Store — 4.4m²
- 10 Boiler room/ Emergency Generator — 26.4m²
- 11 Main Entrance with Air-Lock — 5.7m²
- 12 Children's Toilets — 6.9m²
- 16 Main Stairs — 8.5m²
- 25 External Garden Store — 11.9m²

FIRST FLOOR:
- 04 Cloaks / Outdoor Clothes and Boots — 8.5m²
- 05 Cloaks / Outdoor Clothes and Boots — 3.8m²
- 05-S Kindergarten Classrooms / Activity Area with Sleeping — 69.0m²
- 10 Cleaner's Store — 5.8m²
- 12 Children's Toilets — 9.7m²
- 12 Children's Toilets — 9.9m²
- 13 Resource / Training Room with Presentation Facility — 77.6m²
- 14 Parent - Infant Programme Room Discovery Centre for 1 - 2 Years — 55.9m²
- 15 Balcony — 37.5m²
- 15 External Balcony and Means of Escape — 17.4m²
- 16 Main Stairs — 9.7m²
- 17 Adults WC — 4.0m²
- 17 Adults WC — 2.8m²
- 18 Baby Change — 2.7m²
- 19 Head of Centre Office / Meeting Room — 22.5m²
- 20 Void Over Entrance 'Mini Piazza' — 56.2m²
- 21 Laundry — 22.4m²

First Floor Plan
scale 1:100

Ground Floor Plan
scale 1:100

N ↓

Proposed Floor Plans
scale 1:100
PRESENTATION: 09.11.2010   DWG NO: 1989/01 Rev.5

Jurgow Park Kindergarten proposed floor plan.

153

Mark Dudek Associates
education design

Section A-A
scale 1:100

Section B-B
scale 1:100

Roof Plan
scale 1:100

Roof Plan & Sections
scale 1:100
PRESENTATION: 09.11.2010   **DWG NO:** 1989/02 Rev.3

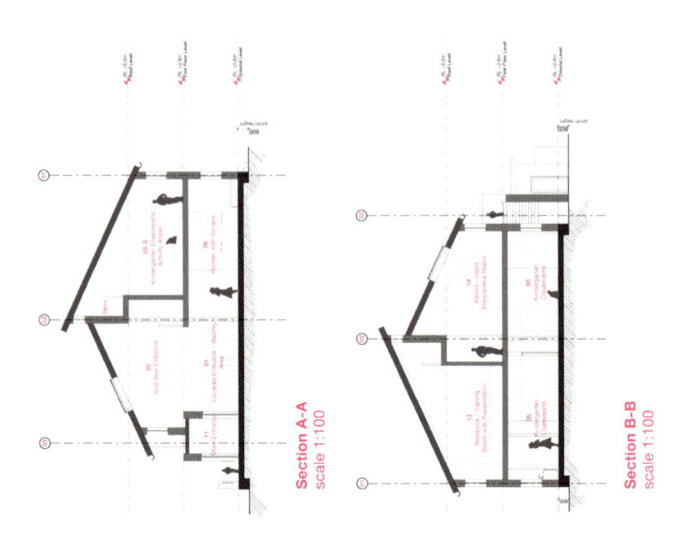

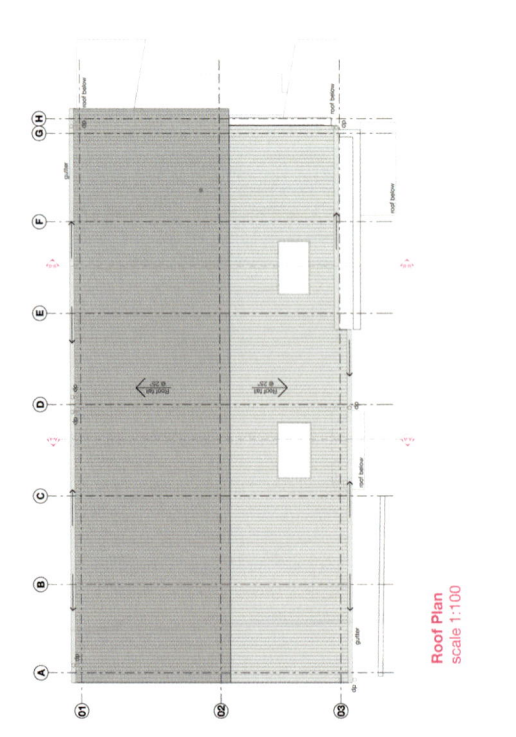

The Jurgow Nursery with the split section creating a complex spatial dynamic.

154

Mark Dudek Associates
education design

view 1

view 2

view 3

view 4

3D Views
scale n.t.s
PRESENTATION: 09.11.2010   DWG NO: 1989/04 Rev.3

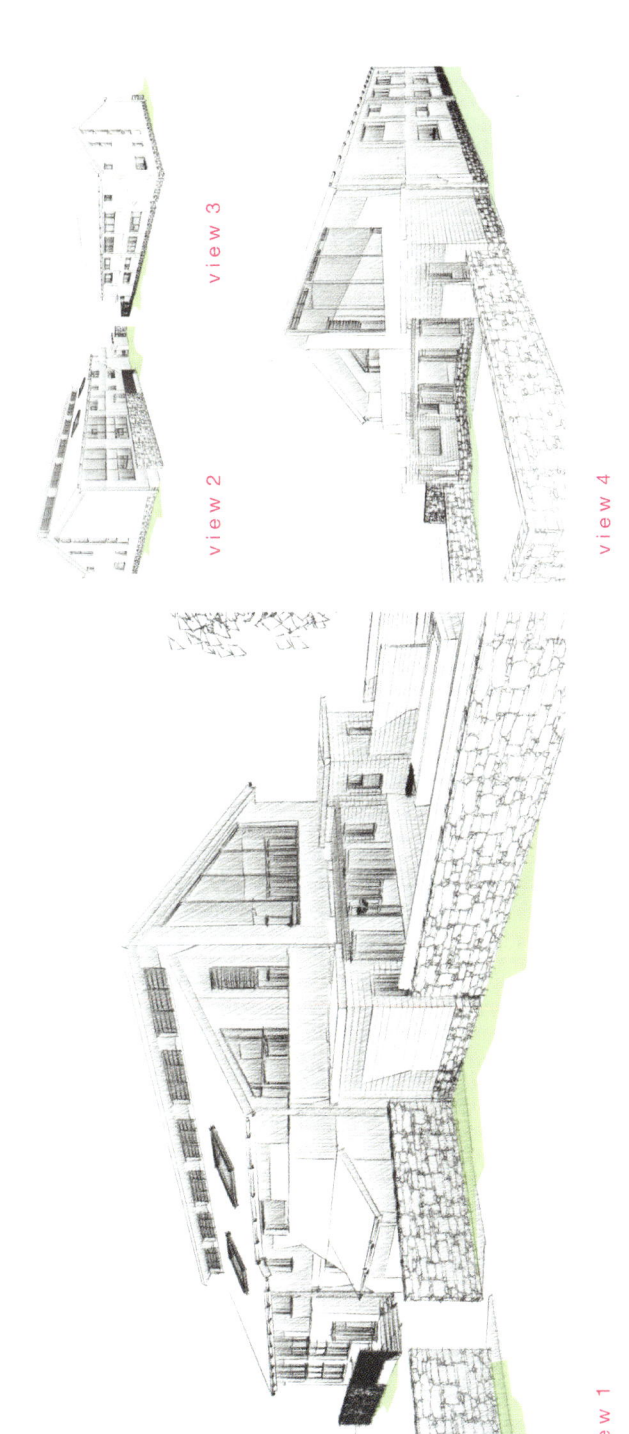

The Jurgow Nursery with big windows affording nourishing green views towards the park (views 1 and 4).

A nursery brief

to the west façade is wide and will provide good, safe short-stay parking for parents dropping off or collecting their children by car.

The fortunate proximity of Jurgow Park, which is adjacent to the site, together with the magnificent views of the surrounding mountains, presents the opportunity to frame this outlook in the form of large picture windows on the south façade. Famously, environmental psychologist Roger Ulrich found through his research that well-being and recovery rates in (adult) hospital patients allocated to wards with a view over parkland (as opposed to a dull view overlooking other hospital buildings), improved more quickly.[3] Not only that, patients' self-reports of their well-being and contentment were far more positive than those who had no view. Green views were good for them, therefore the decision to locate four of the five main children's rooms on this main façade oriented to the park was based on well-founded research around the nurturing aspect of the 'green view'.

Normally nurseries and kindergartens will be located on generous sites where the children will have their own dedicated and secure outside play spaces. However, in this case the site is confined and the available outside play space is restricted to an area of only 35 m^2 on the south side of the plot. Given that there may be at least 72 children on-site at any one time, this gives a ratio of half a metre per child of usable secure outside play space. The ratios recommended by European guidelines is in the order of 9 m^2 per child. However, it is often the case in urban situations like this that children will be able to use nearby public parks for their outside play activities, accompanied at all times by adult carers. The sight of groups of toddlers walking down city streets tethered securely together by purpose-made reins is commonplace, particularly in France and Spain. Here, Jurgow Park provides the ideal situation for the controlled use of the existing play areas within the adjacent park.

Playing outdoors is key to developing young children's bodies and minds and, according to children themselves, it is what they like best at kindergarten. Easy run-in–run-out access between indoors and outdoors extends the child's field of learning, and can help them develop an appreciation of the natural world. In the best possible environment, access doors lead directly from inside onto gardens. A level threshold encourages circulation between the indoor areas and protected external areas, allowing equipment to be wheeled or carried outside so that activities can continue in both environments and children feel free to move between the two.

At Jurgow Park Nursery, the internal floor level is approximately 600 mm above the external ground level. The two main activity areas which overlook the external space will therefore be provided with a level threshold by way of an elevated timber board-walk plinth with a sunken sand-pit in the centre. This level gradually steps down, providing 'affordances' for the children with the use of a variety of floor finishes – stone, timber and rubber compound, for example. Small areas of planting are provided for the cultivation of fruit and vegetables, with stepping stones that enhance the child's sense of their proximity to nature. A raised platform with a mini-den constructed along two sides of the existing walls to Jurgow Park, one metre above ground level, provides views into the park and den areas beneath. There will be an area for water play. Secure, well-designed activity corners provide opportunities for

A nursery brief

View across the courtyard at the Grantham Children's Centre, illustrating the need for a level threshold between the inside and the outside to aid run-in–run-out play.

children to 'withdraw' from the main play space and develop their own sense of community with other children. Unfortunately there is little space for challenging gross motor activities and no climbing frame will be included. Activities will be limited to small motor play, with different materials such as sand, water, soil (for digging) and timber blocks. Jurgow Park itself will provide areas for running, jumping and climbing. Nevertheless, this space will be a valuable and complementary asset to the internal learning environment.

The floor is the main activity surface that young children mostly use, although it is fair to say that children under the age of about two years

In this Paris nursery, polythene shoe covers are provided for all visitors. Staff and children wear indoor shoes and change whenever they leave the building to avoid bringing contaminates back in – the floor is the young child's main surface for activity. It should be treated as if it was as hegemonic as a table top for eating breakfast.

157

A nursery brief

will begin to use child-sized tables and chairs for some activities. Nevertheless, the floor will remain the main place for play. We will explain the thinking on the floor finishes, and the provision of raised plinths and sensory materials in our home base/activity area sections. Developers and managers should bear in mind that hygiene is important and protecting the floors from potentially dangerous bacterial matter brought in on the soles of shoes is important. The removal of all outdoor shoes will be rigorously enforced, with the provision of plastic hygiene shoe covers if required.

The kitchen is an important part of any early-years environment. It symbolizes the homely quality of meal preparation common to any social group. Furthermore, a readiness to provide good, wholesome food and a well-balanced diet for growing children is integral to good practice. We therefore provide a full catering kitchen. It is always useful to involve parents in planning menus. The layout of the areas around the kitchen should encourage

A full catering kitchen is required; this kitchen, at the Familio Children's Centre, Mansfield, includes a washing machine and a full-size dishwasher. Evidently it is short on useable preparation surfaces. This kindergarten in Frankfurt features a window which reads as the letter 'K' from the garden, to illustrate to children its location – architecture speaks to the child.

staff to sit with children, the grouping of children for mealtimes at dedicated tables if possible, encouraging independence, choice and social interaction; thus we meet the needs of children, especially those who arrive early in the morning and who leave late in the afternoon. For them the nursery will become their 'home from home'.

Area guidelines

The ideal size of a typical nursery is dependent on the best group sizes (from a child's and from a staffing perspective) and the need to provide care and education across the age range, usually from new-born babies to young children aged five or six years). At Jurgow Park Kindergarten the requirement for baby daycare is not significant right now, although a crèche-type provision will be offered and catered for in the parent–infant room and in the training room, which will provide two-hour baby care for parents in skills training. The two rooms are co-located side by side on the first floor, overlooking the park.

The younger the children, the smaller the group size should be. The other factors which determine group size are the required staffing ratios and the area of each home base. Generally, the larger a nursery is, the more effective it is in terms of the deployment of staff and the availability of specialist staff resource equipment. Care must be taken to balance group sizes in order to give children the security of working with small and consistent groups of staff and to avoid a large 'institutionalized' feel.

The requirements of the local registration authority will establish space standards and the extent of fixtures and fittings. In the United Kingdom these are now governed by a national standard in accordance with the UK Children Act (1989), ratified by Parliament. In addition, we would expect the home base/activity areas to be registered, increasing the capacity of the nursery to a maximum extra in line with the space standards if required at a later date. Also, the recommendations of the local environmental health department and the fire officer must be incorporated during the development of the project. For example, some environmental health departments do not like open children's toilets within the activity spaces, especially when these areas are also being used for eating. This example is very much a local decision rather than a hard-and-fast rule. Other health and safety issues may be raised, such as the need for a level threshold running between the inside and the outside areas, which may be considered an essential prerequisite for child safety. This, in turn, imposes detailed design requirements for any child-accessible doors leading to outside play areas.

The best space allocation is approximately 18 per cent over the legal minimum as specified by the UK Children Act. I feel that the extra space provides better-quality play and accommodates furniture without restricting movement too much. Managers must be careful to avoid too much furniture and clutter in the home base areas. Specific equipment and furniture required would include children's play zones with child-scaled discovery areas such as mini kitchen areas and play houses, together with some adult furniture. It is good to include a couple of sofas for adults at strategic points within each home base. The standard space quoted is clear activity area space, excluding any fixed furniture or cupboards but including moveable furniture:

A nursery brief

Amy sitting in her 2.3 m² of space, the minimum allowed in her nursery.

Age: 0–2 years   3.7 m² per child
Age: 2–3 years   2.8 m² per child
Age: 3–6 years   2.3 m² per child.

In addition, there should only ever be a maximum of 26 children in a single room for obvious 'crowd control' reasons. With many children in one room the acoustic performance of these spaces is critical, and sound-deadening surfaces, varied ceiling and wall panels, will help. Acoustic ceiling and wall panels are now available commercially. All aspects of the design must comply with local planning permission requirements and building regulations and all other local and national statutory body requirements. Standards relating to other areas such as children's and staff toilets will be dealt with in the appropriate sections below.[4]

Adjacency diagrams

When the initial rooms list has been agreed, together with the basic areas required for each room on the list, it is useful to produce diagrams which show each major space and other subsidiary rooms to which they relate; we call these 'adjacency diagrams'. So, for example, a main activity area will require children's toilets and cloakroom areas close by or immediately accessible. This theory can be presented as a scaled, abstracted diagram which will enable informed discussions to take place prior to the design commencing in earnest.

    It is a profound truth that in order to be a good early-years building, a good client is required. A good client is one who communicates clearly with the design architect, building a bridge between the users and the builders. They should be able to explain where things should be and more importantly, why. It is important to provide as clear a brief as possible in order for the

05/09/2010
Schedule of Accommodation / Adjacencies Diagram – Option 1
To be read in conjunction with Generic Brief (Doc. 1Z.)

1/ Parent-Infant Programme Room / Discovery Centre for 0-2 years. (50 sq M.)
2/ Small Hall / Central Piazza. (20% of unit area).
3/ Resource / Training Room, with presentation and seating facilities. (80 sq M.)
4/ Cloaks / Outdoor Clothes and Boots, outside each Activity Area. (5 x 8 sq M.)
5/ Kindergarten Classrooms / Activity Areas, ECD, 2-3 years, 3-4 years, 4-5years. (2 x 70 sq M.)
5-S/ Kindergarten Classroom with dedicated sleeping function and storage. (70 sq M.)
6/ Observation Room with 1 way mirror for training purposes orientated into both Kindergarten Classrooms. (20 sq M.)
7/ Internal Class Storage, 800 mm deep full-height cpbds. (5 x 4 sq M.)
8/ Medical Room, adjacent or in the same space as Sec + Admin. (12 sq M.)
9/ Secretary and Admin. Also acts as security control access to main entrance area and reception desk. (12 sq M.)
10/ Head of Centre, Office + Meeting Room. (20 sq M.)
11/ Staff Room, with dining table, sofas and soft seating, also requires sink, oven + micro-wave, fridge and kitchen storage areas. (30 sq M.)
12/ Kitchen and Servery, requires full range of kitchen equipment with walk-in pantry. (30 sq M.)
13/ Laundry Room. (12 sq M.)
14/ Staff Store, secure, adjacent to Staff-room. (8 sq M.)
15/ Wheelchair Accessible W.C. (3.3 sq M.)
16/ Cleaner's Store, requires mop sink and shelving. (10 sq M.)
17/ Children's Toilets. (3 x 12 sq M.)
18/ Baby Change, relates to Discovery Centre. (6 sq M.)
19/ Adult WC, separate male + female each with ventilated lobby. (2 x 1500 sq M.)
20/ Outside Covered Play Area. (full width of each external classroom wall and 1200mm deep).
21/ Covered Waiting Area and Entrance with reception desk. (10 sq M. minimum)
22/ External Garden Store, may be part of covered outside play areas.. (20 sq M.)

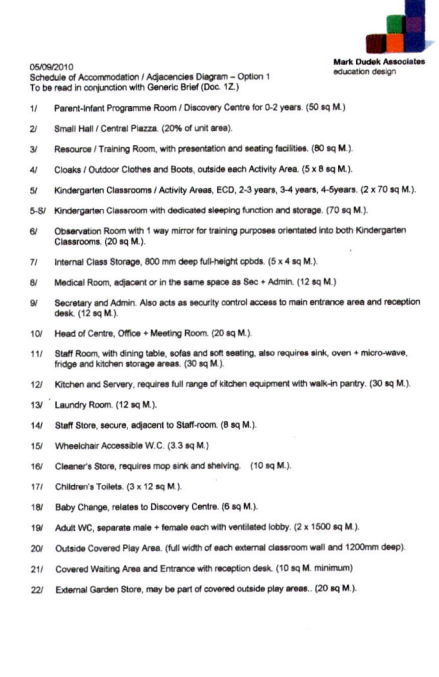

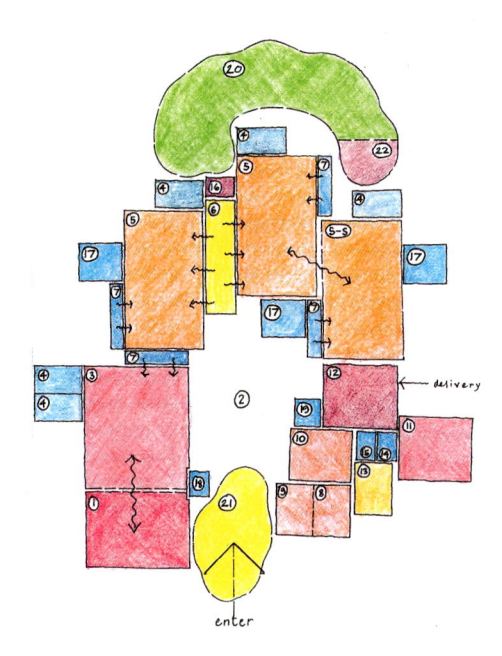

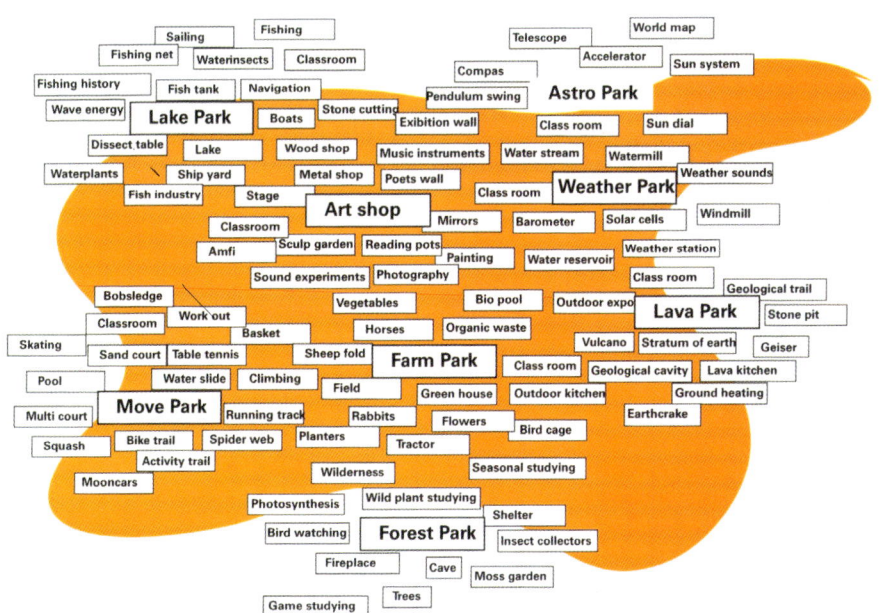

Adjacency diagram relating to early-stage development proposals for Jurgow Park Kindergarten. The arrows indicate direct connections required between different rooms. The semi-abstract nature of the presentation enables ready understanding for some who find architect's plans overly complex and difficult to understand (Mark Dudek Architects). By comparison, the themes set out in the second diagram consist of a large number of programmatic possibilities organized loosely around existing landscape features; this makes the briefing document and the designer's responses deliberately complex, yet open to a range of different interpretations in this early-stage document for the Sustainable School at Urridaholtskoli, Iceland by architects Arkitema.

A nursery brief

architect to initiate the first site-specific designs. This early-stage work will help to clarify issues which are highest priority and those which are secondary. In practice, an informed designer will subsequently discuss the proposals regularly as they emerge in drawn form, refining and improving them as they evolve. Architectural issues will then merge seamlessly with pedagogic priorities. He or she requires an informed individual or working group to consult with during this time. However, time pressures invariably squeeze the consultation process, and even the most thorough pre-construction briefing will fail to iron out all the snags; it should be anticipated that changes will be an inevitable process of user testing after the building has been completed.

In relation to the development of the Jurgow Park Nursery, initially a staffroom and a child observation room were included in the schedule and are shown on the adjacency diagram. Subsequently, when the planning was made to fit on the available site area, and to function as a two-storey building, these spaces were omitted as the client was keen to emphasize the spatial openness of the interior and to enhance architectural legibility generally. This priority only emerged as a function of the architect's first-stage design drawings.

The final schedule of accommodation (or the rooms list) shown below was taken from and measured off the architect's first to-scale plans, so the areas do not correspond precisely to those indicated on the preliminary rooms list. For example, the original areas specified for each home base/activity area was $70\,m^2$. Here they range from $67-70\,m^2$. The schedule presented below may still be adjusted, but it is substantially correct right now. It should be read in conjunction with the related adjacency diagrams.

—— Jurgow Park Nursery – Schedule of Accommodation / Rooms List ——

Ground floor

1. Parent–infant room. Requires 'discovery' area for 0–2 years. Also requires baby change facility and children's toilets. ($60\,m^2$)
2. Small hall/central piazza.
3. Training room. Requires some storage and moveable stack-away seating, projection wall for presentations, coffee and tea with sink and cupboards, four computer work stations and 'discovery' play zones. Children's toilets and baby change must be directly accessible. ($70\,m^2$)
4. Cloaks/outdoor clothes and boots, outside each activity area adjacent to the children's play-yard. ($6\,m^2$)
5. Home base/activity areas. ECD, 3–4 years, 24 children in each. In-built storage for resources comprises three tiers of spur shelving 500 mm deep and 1,000 mm or 600 mm wide. Requires direct access to secure garden/outside play and direct access to internal children's toilets. Children's cloakrooms should be adjacent to the external children's play area. Will have dedicated 'wet-play' specification. ($67\,m^2$)

5-S Home base/activity areas. ECD, 2–3 years, 24 children with in-built storage for resources. Requires sleeping facilities with fold-away beds. ($68\,m^2$)

6 Observation room with one-way mirror for training purposes into both kindergarten classrooms.
7 Internal class storage, 800 mm deep, full-height cupboards.
8 Medical room. Adjacent to or in the same space as secretary and administration. (6.4 m^2)
9 Secretary and administration. Also acts as security control for access to main entrance area with vision panel looking out to entrance gate and inside to the mini piazza airlock. Also functions as the medical room for sick children and should include a screened off area with a child-bed. (16 m^2)
10 Head of centre office/meeting room. Secure storage of family records, sink and worktop with cupboards for tea- and coffee-making facilities. Internal views into mini piazza below. (32 m^2)
11 Staffroom with dining table, sofas and soft seating; also requires a sink, oven, microwave, fridge and kitchen storage. (30 m^2)
12 Kitchen and servery. Requires full range of kitchen equipment with walk-in pantry, including double fridge, dishwasher, oven and storage above and below. Refer to detailed specification below. (20 m^2)
13 Laundry with washing machine and dryer. (3 m^2)
14 Secure staff storage, adjacent to staffroom.
15 Wheelchair-accessible WC.
16 Cleaner's store.
17 Children's toilets with vision panel for observation from activity areas. (Two at 9.7 m^2)
18 Baby change. Directly accessible from training room. Refer to detailed specification below. (2.5 m^2)
19 Adult WC, separate male and female. Adult WC, wall-mounted hand-wash basin at adult height with chrome pillar taps, hot air hand dryer, soap dispenser and towel rail. (4.0 m^2)
20 Covered outside areas adjacent to the main entrance with fixed adult seating and area for the storage of buggies. Fixed bar for locking buggies to be left outside during the day.
21 Covered waiting area and entrance (with airlock), the 'mini piazza'. Includes reception desk, timber bench (with gently curved tapering edge approximately 500 mm deep, 3,000 mm long, 400 mm high), and noticeboards with 'stage' storage feature. Will have direct access to the administration office, toy store, the kitchen servery, accessible WC, main staircase and children's activity areas. This area has a double-height volume above the entrance. It acts variously as a café, community meeting hall, a market and a children's playground. (160 m^2, 20 per cent minimum of net area)
22 External garden store, may be part of covered play areas.

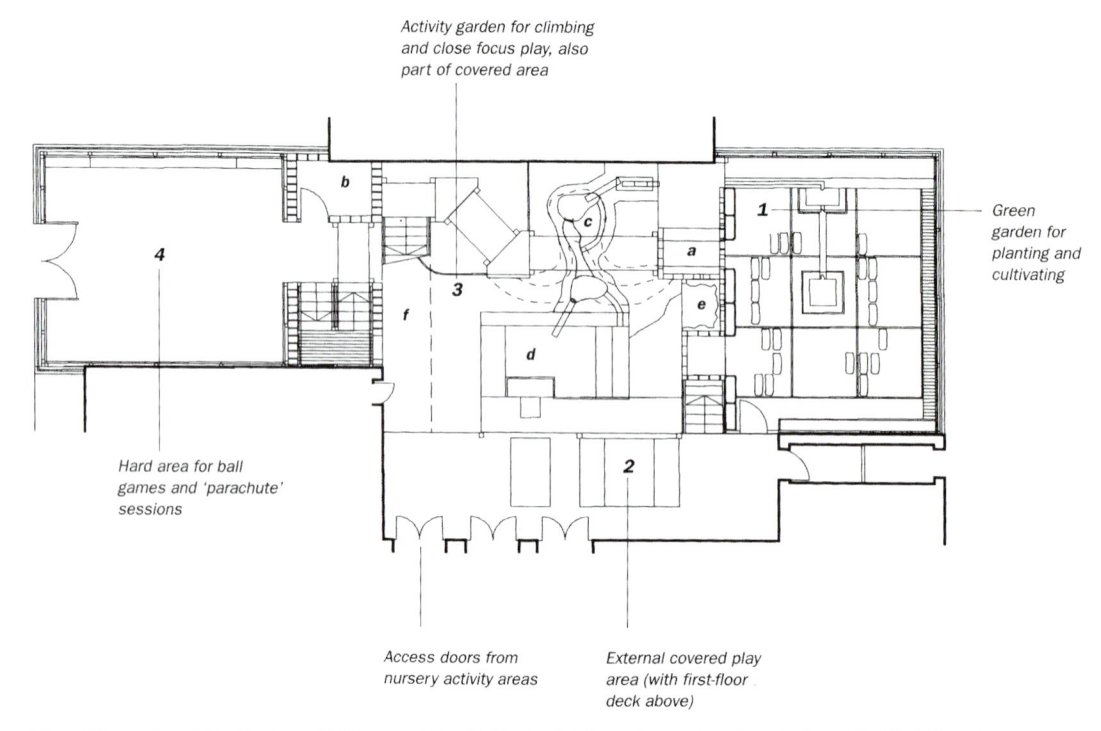

A nursery brief

Activity garden for climbing and close focus play, also part of covered area

Green garden for planting and cultivating

Hard area for ball games and 'parachute' sessions

Access doors from nursery activity areas

External covered play area (with first-floor deck above)

Plan. This garden at the Portman Children and Family Centre in Westminster, London, designed by Mark Dudek Architects, makes optimal use of the available outside space by creating a number of different garden 'rooms'; each is intensively designed to scales of small children to significantly enhance the experience of play.

Key: 1, green garden for planting and cultivation; 2, external covered play area with security roller shutters for safe evening storage of toys; 3, activity garden for close focus play with interactive water feature; 4, hard area for ball games and 'parachute' sessions; a, secret tunnel; b, music deck; c, water activities; d, shallow pool; e, children's den; f, staircase up to the first-floor deck.

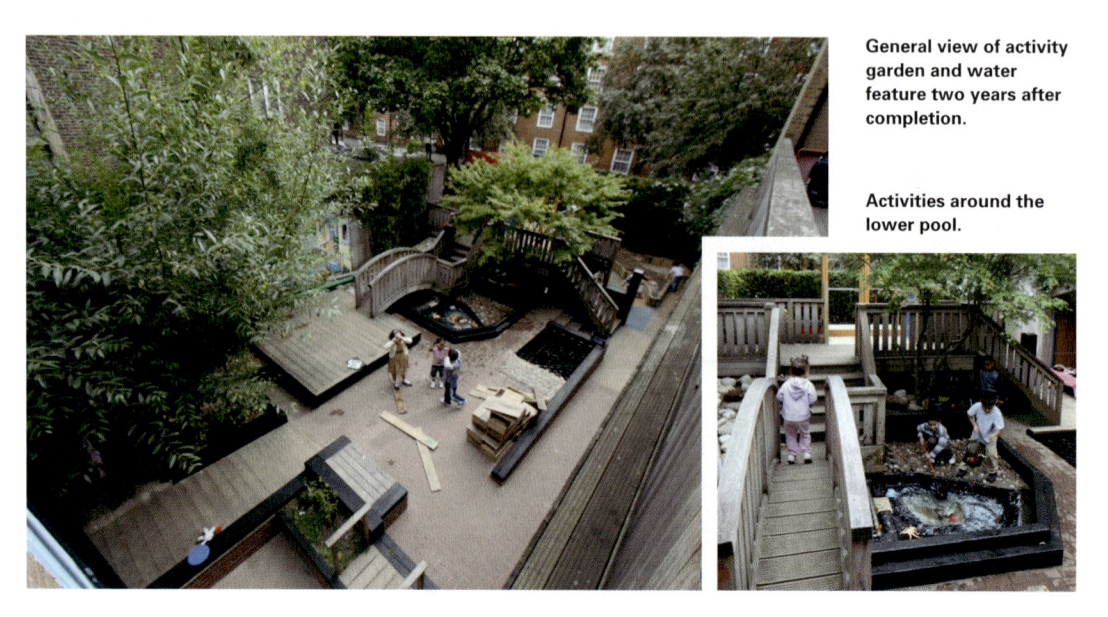

General view of activity garden and water feature two years after completion.

Activities around the lower pool.

A nursery brief

The City of London mural, a theatrical backdrop to the stage/performance area.

Activities around the upper pool (the boy is channelling water down the pipe).

A nursery brief

First floor (see plans p. 153)

4   Cloaks/outdoor clothes and boots. (Five at 8 m^2)

5   5-S home base/activity areas. ECD, 2–3 years, 24 children with in-built storage for resources. Requires sleeping facilities with fold-away beds. (68 m^2)

14   Children's toilets with vision panel for observation from activity areas. (Two at 9.7 m^2)

13   Training room. Requires some storage and moveable stack-away seating, projection wall for presentations, coffee and tea with sink and cupboards, four computer work stations and 'discovery' play zones. Children's toilets and baby change must be directly accessible. (70 m^2)

14   Parent–infant room. Requires 'discovery' area for 0–2 years. Also requires baby change facility and children's toilets. (60 m^2)

15   Balcony above mini piazza with feature seat, double-height handrails to balustrade with child-height vision panels, low-level safety lights. (36 m^2)

16   External balcony and external stair as alternative means of escape from training room and parent–infant programme room. Requires child-safe gate at top of stair. Stair should be minimum 1,200 mm wide. (17 m^2)

17   Main stairs. Balustrades with vision panels, good lighting including low-level safety lighting, double-height handrail, visibility strips to treads, tread sizes to have 290 mm tread and 170 mm risers max. (10 m^2)

18   Adult WC, female. Adult WC, wall-mounted hand-wash basin at adult height with chrome pillar taps, hot-air hand dryer, soap dispenser, towel rail. (4.3 m^2)

19   Adult WC, male. Adult WC, wall-mounted hand-wash basin at adult height with chrome pillar taps, hot-air hand dryer, soap dispenser and towel rail. (4.0 m^2)

20   Baby change. Directly accessible from training room. Refer to detailed specification below. (2.5 m^2)

21   Head of centre office/meeting room. Secure storage of family records, sink and worktop with cupboards for tea- and coffee-making facilities. Internal views into mini piazza below. (32 m^2)

22   Void area over mini piazza. (50 m^2)

—— Design Notes on the Main Spaces ——

Baby home bases

Although there are no baby home bases for daycare currently included in the proposed Jurgow Park Nursery, we include this brief description for reference.

The main home base/activity area will usually be designed for nine babies aged from four months up to two years. This assumes one carer per three babies. The space will primarily be laid out as floor area, with sofas and soft furnishing for carers and for soon-to-be-crawling babies. Low-level windows affording views outside for mobile babies is helpful. The emphasis will be on a quiet, calm environment with good sound-reducing acoustics. You

A nursery brief

might describe it as a 'soft' space, with natural warm materials and calm gentle colours. Soft music, ceiling hangings and wall mirrors will add to the ambience. In reality, young children seem to grow so quickly that confinement in their own home base is not recommended. Even very young babies enjoy the social benefits of mixing with older children within the framework of close care, as illustrated earlier in the image of a Paris kindergarten, where carers move around with 'their children' spending time with other carers in toddler and pre-school areas containing differently aged children (see page 157).

A separate dedicated sleep area is required with one cot per baby. This room would have to have a vision panel and a listening device so that carers can hear and see when children are disturbed. Low lighting and good ventilation should provide an ideal environment for sleep, with higher than average ambient temperatures. Very young children should not have to sleep within the home bases.

For safety reasons the position of low-surface-temperature radiators or underfloor heating must be carefully considered. No carpet should be specified; instead, and if affordable, a non-slip wooden floor works very well. Loose rugs which can be removed for cleaning or replaced if too soiled are ideal, with a vinyl or a linoleum floor finish throughout if a wooden floor is not affordable.

A milk kitchen designed as a separate small unit directly accessible from the activity space is required. This will usually contain cupboards, shelving, a sink with drainer, a small fridge and microwave. The two-plus age group do not need access to their own dedicated milk kitchen. None is required for the pre-school either – they would instead use the main kitchen. However, drinks and snacks will be delivered to the home base at certain times of the day.

A nappy-change table must be provided, ideally in its own separate ventilated room. The nappy-change table must have a raised lip of at least 7 cm to prevent the changing mat slipping off. The height of the change unit is normal kitchen-worktop height (900 mm). Shelving above the changing area for nappies and creams should be provided. Separate shelving niches are

A 'Loxos'-type baby-change unit with padded moulded top and a baby bath beneath. A safe and hygienic, if expensive, unit which also requires a lot of free space. This purpose-made corner unit organizes everything needed within a relatively confined area. It does not work very well for left-handed people; it has a soft-base cushioned mat, raised edge for safety, sink, soap dispenser (out of sight) and baby-wipe dispenser all in close ergonomic proximity. Note that mirrors provide distractions, with head height storage for individual baby's change items.

A nursery brief

required with baskets for individual children's belongings close by. A paper towel dispenser located within easy reach of the changing surface is important. The French-made Concorde or Loxos tables with built-in baby wash bowls are the best, although they are expensive. Other requirements are:

- a low sink next to the changing table – the sink should be to the right of the changing area;
- a hand-wash basin with taps and soap dispenser sited close to but away from the nappy-change unit;
- a noticeboard in the changing area;
- finishes should be hygienic and the wall be tiled to at least a four-tile height above the changing worktop;
- mirrors fitted to a lowered ceiling and surrounding walls helps to distract babies;
- a vision panel from the main activity area is required if baby changing is in a separate room;
- extraction ventilation or direct window ventilation is required;
- hygienic dedicated waste bins for nappy disposal to be positioned close by – appropriate refuse disposal generally must be discussed and agreed with the client;
- baby-change storage pods above for personal possessions, such as medications, soft toys and 'cuddlies'.

The following other facilities must be provided:

1. A child WC directly accessible or close by, possibly from each half of the activity areas if paired, with a nappy-changing worktop (with raised lip) for older babies. Total area to be $3–4\,m^2$ minimum. This area should contain:

   - two 'Junia-12' toilets – plumbed-in 'potties' (alternative make may be specified)
   - two child hand-wash basins – note the need for thermostatically controlled child-safe taps.

2. Easy access to a milk kitchen from the younger toddler home bases, primarily designed to provide for simple baby needs, with shelving, sink, small fridge and microwave. Depending on the size of the centre, one or two milk kitchens within the building are usually required unless there is a central full-catering kitchen. These do not need to be within or accessed directly from the single or paired home bases.
3. If the milk kitchen is not accessed directly from a pair of home bases, then there needs to be an area within the pair where crockery and catering equipment can be placed for collection, where meals can be delivered to and served from and where bottles can be stored. This area should have a sink and worktop with a small fridge. May be shared with baby unit milk kitchen.
4. Additional built-in storage for nappies and babies' other personal items required near the entrance door.
5. Cloaks and other personal storage for staff, which can be built within the room rather than in a separate staff area if possible.

A pair of accessible sinks set into the timber worktop, end-on, at this art area for 3–4-year-old children at the Stanley Infant and Nursery School (designed by Mark Dudek Architects).

6   Low-level sink to be fitted into one half of the pair of home bases for use by children washing paints, utensils, etc. Storage above and below. This would be used for a number of different purposes.

Toddler home bases for 2–3-year-olds

This home base will be similar to the baby home base in ambience, except it will have fully accessible children's toilets and baby change as described above. Children aged two-plus will begin to become potty trained, although they may require assistance up to age five. These two facilities may be shared if co-located between two activity areas or in good proximity. At Jurgow there is no baby unit, and distinct facilities must be provided, with a cloakroom and a children's toilet with a nappy change facility within as described previously.

A nursery brief

However, no milk kitchen is required as the main kitchen at ground-floor level will provide milk and snacks which will generally be carried by hand to the home base. There needs to be an area within the home base where crockery and catering equipment can be placed for delivery and collection and where meals can be delivered to and served from (including bottles).

This delivery point should have a sink and a worktop with storage beneath, with a small fridge if possible. It may be located close to the entrance and children's cloakroom/storage. For main meals it is good for these older babies to use the communal eating areas within the mini piazza if feasible. Alternatively, meals can be served at tables within the activity space, depending on the pedagogic principles adopted. At the excellent Italian childcare facilities in Reggio Emilia, mealtimes are an important socializing activity, and all early-years children will eat together, including small babies. However, it is worth pointing out that young children require frequent drinks and snacks between the main communal group mealtime sessions, which will generally be provided within the dedicated group activity areas.

The space itself will primarily be laid out as floor area, with sofas and soft furnishing for carers with 'discovery' areas for toddlers. Low-level windows affording views outside are always beneficial. The emphasis will be on a well-ordered environment with good sound-reducing acoustics and, bearing in mind the number of children, capacious storage for resources, both adult- and child-accessible. It is a 'soft' space, with the use of natural warm materials and calm, gentle colours suitable for young children; however, it is more open and varied than the baby room described above, with a range of niche-type seats and discovery-type play areas. It will have soft cushions and play mats to enable children to fall and jump, providing slightly more advanced opportunities for physical play. By this age children will be using child-scaled tables and chairs for some of their activities, although for the majority of the time they will be using the floor as their main activity mode.

It should be borne in mind that 24 toddlers have far greater need for movement and physical activity than nine babies, particularly male toddlers. Confinement in their own home base for long periods of time is not recommended. Hence the use of additional play zones such as the mini piazza (for use during very cold winter periods), the children's courtyard garden and Jurgow Park itself. However, there will be a lot of demand from other children for these areas. During winter months the mini piazza will be a key additional zone to take the strain off the home base area when it has a full complement of children. Programming the use of these additional alternative spaces is a very important part of the manager's coordinating role, optimizing the use of confined space within the kindergarten, enabling teachers and care workers to concentrate on working with children, rather than managing the space.

Sleeping will be required for full daycare children at this age, usually at 1 p.m. after the midday meal, for an hour or more. However, there will not be enough space for a dedicated bedroom in this particular project. Instead, sleeping mats or lightweight mattresses should be provided along with pillows and bedding which can be laid out on the floor of the activity areas. When not in use they should be stored away for hygiene reasons in well-ventilated cupboards.

A nursery brief

At Jurgow Park the toddler home base is on the first floor and towards the rear of the plan for logistical reasons (there is simply not enough space on the ground floor for all of the child-use accommodation to be contained). It is intended to cater for as many as 24 children, in an area of 70 m² minimum, so space will be tight. When some children are ready to sleep, others will not and they will spend time in other areas of the building in order to avoid causing a disturbance. As one concerned colleague observed, these daycare centres seem to be populated by very tired children, because their sleep is so often disturbed. This is an important management issue.

Although the toddler's room is at the rear of the building and on the first floor, it is bright and naturally lit, with views down over the street and into the mini piazza below. It is also conceivable that children will spend part of the day in other activity areas, sometimes mixing with older children. There will need to be appropriate management flexibility to facilitate organized groups of children moving around the building and its surrounding external areas to extend their field of learning.

The following are also required:

1   A directly accessible children's WC and nappy-changing area with a nappy-changing table (with raised lip) and baby sink with storage, all readily accessible. The area is to be about 3–4 m² minimum, and to have three child-sized WC cubicles with three hand-wash basins (at a ratio of one WC for every ten children). It must be well ventilated, with an openable high-level window if possible. The room will have a vision panel affording views in from the main activity areas. This area should contain:

Children's bench seat with dedicated pegs, one for each child aged 3–4 years; the reality is that most, if not all the time, when this facility is in use, coats and bags are on the floor. A good bench seating height for an early-years child aged 2.5–4.5 is between 330 mm and 420 mm. Inevitably, the higher the seat the more the child likes it.

A nursery brief

- three 'Junia-12' toilets – plumbed-in 'potties' (alternative make may be specified);
- three child hand-wash basins – note the need for thermostatically controlled child-safe taps;
- child-specific soap dispenser;
- paper towel dispenser or individual towels, but a hot-air dryer is not recommended.

2   There needs to be an area within the home base where crockery and catering equipment can be placed for collection and where meals can be delivered to and served from and bottles stored. This point of delivery should have a sink and worktop, with storage and a small fridge.
3   Two low-level sinks to be fitted into one half of the pair of home bases for use by children washing paints and utensils; there should be storage above and below.
4   Built-in storage cupboards for resources 2,300 mm high comprising three tiers of spur shelving 500 mm deep and 1,000 mm or 600 mm wide, with double full-height doors.
5   Children's cloakroom – which can be built within the room or in a niche just-off, rather than being in a separate room – is required.
6   Cloaks and other personal storage for staff, which can be built within the room rather than in a separate staff area, will usually be in the form of individual lockable cupboards, 300 mm deep, 350 mm wide, with a top shelf and hooks below. One per staff member assigned to the group; for them it will also be a home base in the absence of a staffroom.
7   No carpet should be specified. Instead, if affordable, a non-slip wooden floor works very well. Loose rugs which can be removed for cleaning or replaced if too soiled are ideal, with a vinyl or a linoleum floor finish throughout if a wood floor is not affordable.

Pre-school home bases for 3–4-year-olds and 4–5-year-olds

At Jurgow there are two main home base/activity areas which are intended to be used in rotation to suit the varied pedagogic programme during each week. One home base for 4–5-year-olds will be designated the arts and crafts wet area, with low-level sinks and work benches for messy activities such as painting, water and sand play. For this group of children it remains their day-to-day home-from-home, with requisite cloakrooms for personal storage and the sense of belonging that implies. The courtyard garden will be a natural extension of this area, and a number of different modulated floor levels may be provided inside to complement the undulating aspect of the external play zone. The other activity area for 3–4-year-olds will be designed as a clean area supporting dry activities which will include computer tables, soft reading corners with books, children's mini play areas such as scaled kitchens and other designated zones for activities such as block play and role play activity. The computers should be built into a designed unit to allow permanent installation with good cable management (safety is an issue). It will be a more cosy area than the wet side, with sofas and a range of soft tumble zones. Vinyl or lino with rugs are required in the dry home base. On the wet side, vinyl or lino throughout or if affordable a fully waterproof wetroom specification with floor drains will radically improve their performance. Tiled wipe-down surfaces are required behind all sinks.

All other facilities will be as above, with children's toilets with a baby-change table and a mini shower. Children's cloakroom areas are designed in the form of lightweight 'sheds' ameliorating between the interior and the garden zone, mini houses in tune with children's scales rather than adult scales.

The courtyard garden

Although the area available is limited to 85 m^2 at Jurgow Park, this is an extremely important part of the kindergarten brief and will be used intensively. It is located on the south-facing side of the building overlooking the public park. However, it will have no direct access to the park; security to the entire building is controlled by the main entrance only. It will be enclosed by a wall/fence that should be 1,800 mm high. Spy holes for children to look out should be incorporated into the design of this enclosure, with no views in.

The courtyard garden will be scheduled for use by various groups throughout the week, as if it was any other internal room. Its scales and sensory arrays are intended to enable it to function as the 'outside classroom' where much learning-through-play will occur. It will be challenging both physically and mentally, and is zoned to accommodate a number of differently themed activity zones and play features which, while intended for child use, must be fully accessible for adults as well. The space should include some or all of the following:

- external covered play areas;
- a house or a cave under cover of the deck;
- a flat, even space for children to use building blocks and to build other constructions;
- nooks and corners with wooden seats or benches where they can read and write;
- integrated child-sized chairs and seats wherever space allows;
- good-quality adult-sized garden benches;
- planting which the children can dig and cultivate themselves;
- a treehouse and/or raised deck;
- a stage with a removable canvas canopy for summer use;
- a sand-pit;
- a music deck;
- a secret tunnel;
- a water play area.

It is intended that within this relatively confined space a truly challenging and diverse outdoor area will be created to provide a range of features to promote sensory play and to extend indoor play beyond the boundaries of the building. The most important aspect of this offer will be the range of different levels across the area, which some (adults) may find hazardous. They are in practice

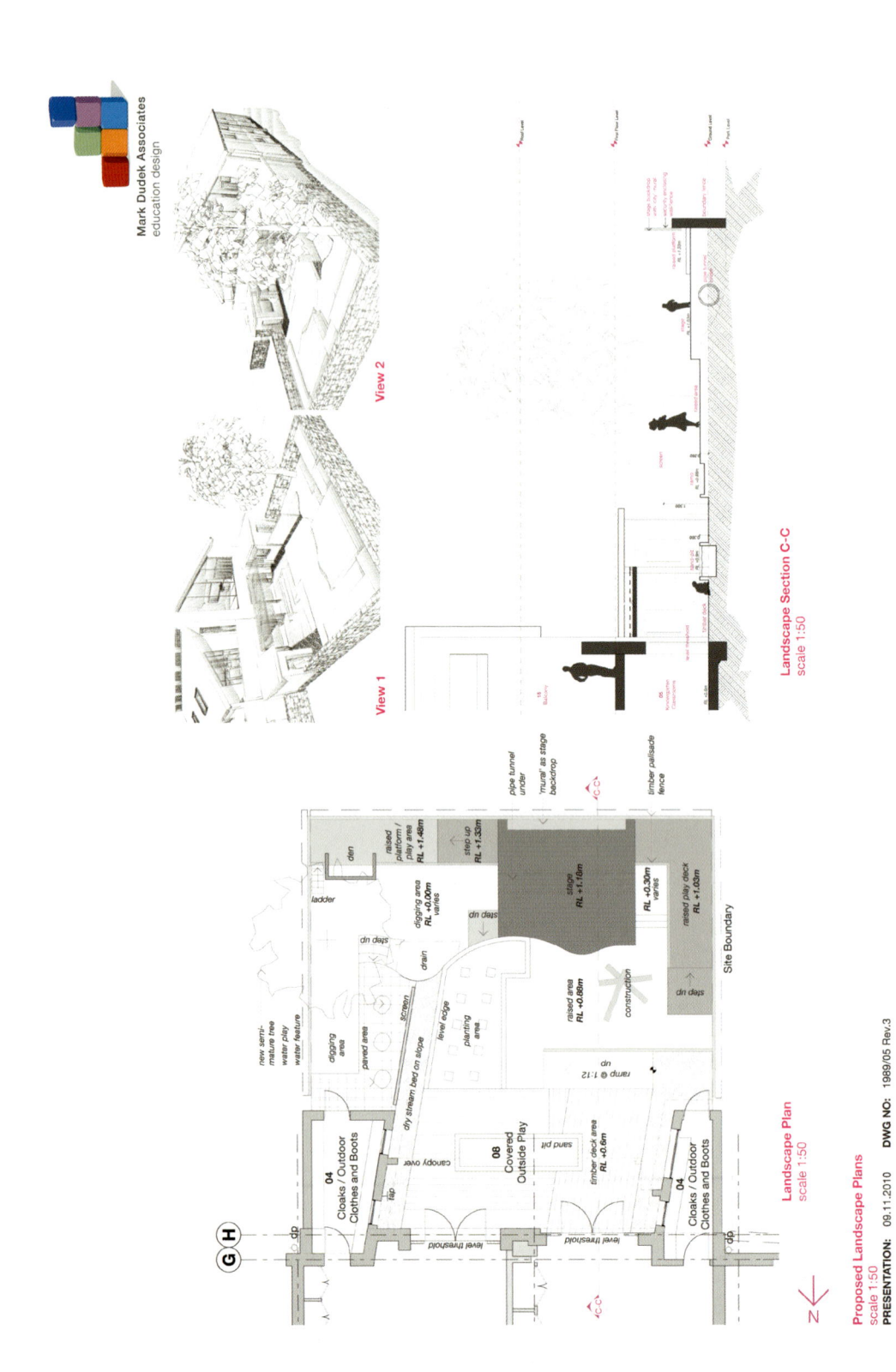

The design for Jurgow Park Nursery garden: Note the manner in which the building form seemingly embraces the garden.

safely negotiated by most children using them. These level changes promote movement among the users developing wayfinding skills, each journey enhanced by a range of sensory stimulations along the route, including different floor and wall textures, planting boxes for herbs and other aromatic plants, wind chimes, gongs, pet enclosures, mirrors and even an anemometer to measure rainfall. This optimizes a limited amount of space and enables children to use all of their senses to explore in a safe, controlled environment. In this particular situation, there are four key levels:

Level 1 (000 mm): the external play deck immediately outside of and level with the internal activity areas. It has warm timber decking and a sand-pit with raised edges formed out of timber.

Level 2 (+515 mm): the intermediate level connected to the external play deck by a gentle ramp; it has a hard, level finish of brick sets. Ideal for peddle cars and construction activities.

Level 3 (+1,330 mm): the stage, an area of play and performance decked out in timber, the natural focus, with a canvas canopy for summer shading.

Level 4 (−520 mm and varies): the cultivation areas with compacted soil and sand.

A rich sequence of activities can be introduced, such as water features; imaginative climbing equipment should also be provided if space allows, but at Jurgow the area is simply too confined. Some limited physical activity will be enabled, with the variations in level and landscape features that children can climb under and over, different textures and finishes with applied colours which promote diversity and movement within the external environment. These devices can all be used to provide an imaginative and ever-varying external landscape. It is an area which will add significantly to the overall experience of the Jurgow Park Nursery.

The mini piazza/playstreet

This is an area through which parents pass every day. It is primarily the entrance threshold. The role of the threshold is to welcome and provide security against intruders. It should therefore be wide and unencumbered so that a double buggy can pass through with ease, but it should also have a reception desk for parents to check-in and receive directions. The head of centre may wish to have an office here to provide a concrete point of contact for parents and to give them somewhere meaningful to check-in. It is also useful if buggies can be stored in the area, although this is not always practical.

In addition to the designated activity spaces it is beneficial to view the entrance area and connecting circulation areas as part of the play zone, a children's area in all but name. At Jurgow this is a large, double-height space around the entrance and kitchen which serves as a circulation and control space, but also as a dining area. At the back the servery feeds directly onto it, just beyond the main staircase and the reception area. However, its major role is as a waiting area for parents. It contains storage for shoes at the front, just within the entrance airlock; it has sofas, noticeboards and leaflets with other

A nursery brief

The Cherry Lane Children's Centre buggy store can be seen here on the far right side of the entrance. It requires a secure bar and lockable chains/ padlocks. It is very rare for any centre to have enough space to contain the buggies inside the centre.

reference material to support family services in the town. Baby buggies should be restricted and left outside if possible.

It is a flexible zone which may at times be used as a play area for children. Therefore soft cushions and small climbing features may be stored close by. It is possible to also include an area to support artistic/creative endeavours so tables and child-height chairs should be included and storage cupboards located close by for resources. Children should have the opportunity to express themselves through movement, drama, music and the arts within this space. At Jurgow there is a stage proposed at the back, a raised area with storage beneath which can be used for theatrical events, with performances by the children. It may also be used for evening events by parents. In the larger kindergartens this area may be split in two, particularly when the pre-school home bases are on the first floor.

This area may also double up as an after-school area. We call it the mini piazza, reflecting the town's full-scale public square soon to be constructed just outside the kindergarten building.

Toilets

The relevant social services department will establish the standard of sanitary accommodation; however, this is usually at the ratio of one WC to every ten children or proportion of ten. The same number of wash basins will be required and both must be of an infant size and at a suitably low level. Half-height doors are required to WC cubicles and pipework should preferably be concealed.

In all toilet areas for under-threes, nappy-change areas need to be fitted. The normal configuration would be:

1. For babies and toddlers up to the age of 18 months, one baby-change table per nine babies, with two plumbed-in potties for older babies.
2. For the under-threes (per pair of home bases), two toilets and one nappy-change table.
3. For the over-threes:
   - two toilets and wash basins for a 16-place facility;
   - three toilets and washbasins for a 24-place facility;
   - four toilets and wash basins for two 16-place home bases.

Separate staff/visitors toilets must be provided as required by building regulations requirements relative to the number of staff employed and to the number of floors in the nursery. If the building has more than one floor level, toilet facilities are required on each floor for staff. Disabled access and facilities are required usually just off the main entrance. The ground-floor toilet may double as a disabled facility and is usually located within the entrance mini piazza.

Kitchen and pantry

Every early-years setting, whether full- or part-time, needs to provide frequent meals and snacks for the children, and so constant food preparation is an essential part of the setting's life. For health and safety reasons, kitchens are often separated from the main children's areas. However, some recent practice illustrates how kitchen areas – which can, to some degree, be integrated into the life of the nursery – will significantly enhance the quality of an early-years setting (see pages 88–9).

The distinction between a real-life kitchen at home and a commercial kitchen in a nursery is that the latter must be non-accessible. Kitchens solely for the preparation of full meals should be out of bounds to younger

A nursery brief

children, but some controlled access for 4–5-year-olds can help them towards a better understanding of the social role of mealtimes in the life of their nursery. Windows and vision panels between children's spaces and the kitchen enable the sometimes theatrical spectacle of meal preparation to be observed in full flow.

Parents and early-years practitioners are increasingly concerned that many young children seem to have poor eating habits, participating in the junk food culture at an early age. This issue has been addressed by many children's centres, which provide only organic food, cooked freshly on the premises. Two Danish centres feature open kitchens where children are encouraged to participate in the preparation of food before it is cooked, in a safely enclosed chef's kitchen.

At Jurgow a full-meal commercial kitchen with a comprehensive range of kitchen equipment requiring an area of at least $25\,m^2$ is needed. This will accommodate storage and worktops necessary for the level of preparation needed. Compliance with local environmental health requirements regarding hand washing is important. This will usually require a separate hand-wash basin with towel dispenser and soap dispenser located away from the paired catering sinks. For a smaller kindergarten the area may be reduced. Food delivery should be directly from a dedicated external access door rather than through the main entrance if possible.

A small servery is required, although the wider the better. At Jurgow this will be the full width of the kitchen and should have a secure roller shutter when it is out of use at evenings and weekends. A walk-in pantry is required, accessed directly from the kitchen. Ventilation is important and a window with an opening mechanism and a cooker/hob overhead mechanical extractor unit is required, transferring stale air to the outside. Facilities within any kitchen must include a fridge, sink and drainers, worktop and storage as a minimum.

Also required are:

- electric sockets as required for a standard commercial kitchen;
- a separate hand-wash basin away from the main work area – a soap dispenser is required over the hand-wash basin;
- double sink and double drainer;
- noticeboard;
- tiles to walls up to ceiling height above all worktops;
- adequate low-level and wall storage cupboards of commercial/industrial quality with closed rear panels and all doors to have stainless steel 180° hinges. Worktops should be 30 mm thick laminated MDF with post-formed rounded leading edges;
- a stainless steel industrial island unit to separate 'clean' from 'dirty' sides of the kitchen. This unit should have storage below and be fitted with an oversize worktop.
- appliances – to be planned for as a minimum are:
    1. commercial six-ring gas cooker;
    2. commercial extractor fan;
    3. commercial larder fridge;
    4. commercial freezer, upright or chest;

5   commercial dishwasher;
6   worktop microwave on or close to stainless steel workstation.

- electrical installation as required to power these items;
- extractor as required by Environmental Health Department;
- a walk-in pantry is advantageous;
- blinds to windows and doors, facing the outside;
- coat hooks in the kitchen (for aprons);
- fly screen fitted to kitchen windows/fly electrocuter is beneficial.

Laundry

A laundry provision must be made separate from the kitchen. It must accommodate as a minimum a washer/dryer and some worktop space, but can be minimal in size – approximately 5 m^2 is acceptable. Additional storage cupboards are useful in this area. A large storage cupboard with shelves for nappies, paper towels and other consumables should be included, space permitting.

Heat generated by the machines is sufficient to heat the laundry room, so no radiators will be required. However, ventilation is important to avoid the effects of condensation. The location of this room should preferably be on an outside wall with an openable window. The laundry room must contain a sink and drainer with cupboards below and shelving above. Electric sockets for the washer and dryer and a mechanical vent for the dryer is required. Hooks for cleaning equipment are also required.

The staffroom

A room of approximately 12–24 m^2 (relative to the size of the nursery), net of built-in storage space, will be adequate for the staffroom. A tea/coffee preparation area and sink unit is required. Space for a fridge is also important. It is preferable to have a separate secure area of approximately 5 m^2 for staff lockers and secure storage. If this is not possible, lockers may be provided within the staff area.

A range of furniture is required, some hard and some soft, with a work table for computers if there is enough space. At Jurgow the client asked that the staffroom be omitted from the brief to provide more space for children. Lockable cupboards for staff belongings should therefore be provided within each activity area.

The concept here is to make this a relaxing escape from the noise and hubbub of the workplace. Consider a green view and even the possibility of a small area of outside space with restful quiet areas for calm contemplation. To some developers this may seem like a luxury, but a small investment in the welfare of staff will pay dividends.

Storage

There is never enough storage. In addition to built-in storage within the activity areas, (with three sets of shelves, 800 mm deep and full height within each activity area), there is a great need for additional ancillary storage space accessible from general circulation areas. For example:

- a toy store 4 m² within the main entrance mini piazza;
- an external toy store relating to the garden, which should be secure and child-safe – this can sometimes be integrated into the covered outside play area and secured with an automated or manual roller shutter;
- an external buggy store which should be within the enclosure but immediately outside the main entrance doors – it should have a security bar and a lightweight covered roof; if space permits, the buggy store should be inside the building;
- a secure cupboard for the storage of chairs and tables to service community meetings;
- a cleaner's store of approximately 8 m² with a sluice area.

Reception and manager's office, medical room

Where space allows, an area which can function as a manager's office and also acts as a subsidiary reception area where parents can leave children in emergencies and where security can be monitored. It is the point where the dissemination of important information to staff and parents is established. The medical room can be included to provide additional space as it will be used only sporadically. At Jurgow there is in addition a head of centre's office located on the first floor.

The manager's office should ideally be accessed from reception. It is an important room for the storage of sensitive records, and should have a small table (in addition to other office facilities), where private meetings can take place. It will need to be secure to protect personal family information. The vision panel providing views onto the entrance area should have blinds or curtains to provide privacy when meetings are ongoing. The office needs to be large enough to accommodate a desk, a filing cabinet, a computer stand, an office chair and an extra chair. Noticeboards and shelving are required. The door must be lockable.

The ideal arrangement is to have the manager's office beyond the reception, where it can be part of the greeting and meeting function, yet so that it can also be closed-off and private on occasion.

External facilities

At Jurgow there is a limited area for dedicated outside play, and the intensely designed courtyard garden will have no conventional physical play equipment for climbing. It is envisaged that Jurgow Park itself will be used by children from the nursery. Where space permits, a much larger external garden/play area is essential, which complements the courtyard garden. It should include enough area for gross motor play. This may be in the order of 1,500 m² or 9 m² per child.

One-third of the area should be hard surfaced for all-weather use. The remainder should have soft safety surfacing and this area may contain suitable, fixed outdoor play equipment such as slides and climbing frames. In addition there should be adequate covered areas to park four or five prams for sleeping babies, in a location which can be easily overlooked and supervised. Like the main area for older children, it should have shading to suit the climatic conditions. It is also very useful to include a separate fenced-off area for babies and early-years toddlers. Also, an external store (preferably part of the building) for larger outdoor play items should be provided. A covered sand-pit is also required. It should have a removable cover for child safety, so it cannot hold rainwater and is resistant to the fouling of small animals at night. Usually it is to be located beneath a canopy close to the activity areas.

Other issues:

- As a minimum, two flagstone widths all around the building should be provided. This helps when practising fire evacuation drills, and aids maintenance generally.
- A gentle slope down from doorways to the ground level if required – no steps if possible.
- Earth moved during construction may be made into grassed mounds for adventure play. The earth mounds could form a small amphitheatre.
- A hard play area (approx. 40 sq. metres) for bikes, peddle cars and ball games will be required with an indeterminate grassed area. There could be a 'kick back' area associated with the hard play. This may be a wall type of construction, or could be timbers driven vertically into the ground in a semi-circular shape.
- The landscaper should endeavour to provide a wide range of trees and shrubs both from an aesthetic point of view and also to promote a knowledge of nature. This should be uncultivated and designed to encourage wildlife, including birds, butterflies and small animals and wild plants and flowers.
- Poisonous or sharp/spiked plants must be avoided (a list of dangerous plants will be available from your Early Years Authority).
- An area should be allocated for the cultivation of fruit and vegetables by the children. A larger allotment section would allow a pensioner to cultivate and encourage children to participate in gardening activities.
- A perfume garden for aromatic plants, especially herbs and spices which can be used in cooking, increases sensory awarness.

Detailed Design Notes

Noticeboards

To be child-visible, noticeboards must be a maximum of 60 cm above the ground and 2 m high. Fitted safety mirrors in some parts of the kindergarten 15 cm above ground level and up to 2 m in height will aid spatial awareness.

Car parks

Parking should be made available for staff and visitors as approved by the local planning department. From a child's viewpoint, the car park must be safe with no concealed entrances or unnecessary reversing.

Curtains and blinds

Slatted blinds, curtains or roman blinds may be fitted. Hygiene and maintenance should be considered. Although curtains may appear to be less hygienic than

A nursery brief

blinds, they can be easily washed and if well designed, curtains will provide a soft, domestic ambience.

Doors

External doors are to be specified through discussion with the fire department and Social Services. Care is to be taken where escape bars are fitted which can be released by children. These should be at high level, with an alarm device fitted.

Home base doors should be sliding or double swing with glazed openings for safety, opening out onto the external play area wherever possible with a level threshold to provide easy run-in–run-out play. Avoid steps or up-stands.

Viewing panels should be provided between circulation areas and internal rooms. Vision panels with safety glazing should be provided at both adult and child height for separate child WC facilities. Doors may be coded by using the shape of window panels to indicate a particular type of space – for example, triangular windows for prohibited areas such as the kitchen and laundry, circular for home bases and square for toilet areas.

Hinges should be protected so that the opening between door and doorframe at the hinged edge is covered by a protective finger guard (of which there are now many on the market) to prevent fingers being trapped. Alternatively, doors are available which use continuous piano hinges which are child safe.

All doors are to be fitted with colour-matched kick plates on both sides. A hand protector fitted adjacent to the door handles is suggested since, as staff frequently push the doors to open with shoulders and elbows the area around the handles quickly becomes scruffy.

The main front door requires an out-of-hours access system and an intercom connected to the main reception. Ideally this should have an automated opening mechanism albeit controlled from an internal reception desk. Self-closing doors need to have spacer mechanisms to prevent the trapping of fingers.

Handles should be sturdy and chunky. Large, plastic lever handles are easily operated and safer than metal. A general principle is that doors through which children must not go unless attended should be fitted with a single handle out of child reach at approximately 1,400 mm height. Dual handles are to be avoided as it is difficult to operate two handles when carrying a child or food or drink. A disabled person would not be able to open a dual-handled door. Door handles from home bases to activity areas should be set at ordinary height. High door handles are to be fitted on doors leading to:

1   kitchen
2   laundry
3   staffroom
4   staff toilet
5   office
6   reception
7   external doors and those to the main entrance.

Dispensers

Soap, paper towel 'spiral rolls', toilet tissues and tissue dispensers are, where possible, to be provided by local suppliers, although the design is to be approved. They should be robust yet child-friendly.

For child toilets the dispensers should be fitted at the following heights:

soap	150 mm above sink
tissue	1,400 mm above floor level
toilet rolls	900 mm above floor level
paper towels	1,400 mm above floor level
hot-air dryer	On the wall adjacent to the sink at a child-accessible height.

Each home base/activity area requires one soap dispenser above the sink, one tissue dispenser on the wall adjacent to the sink and one hot air dryer on the wall adjacent to the sink.

In child toilet areas each toilet unit requires one soap dispenser for every two sinks, one toilet roll dispenser adjacent to each toilet and one hot air dryer per three sinks on the wall near the sinks (but clear of the sinks).

Nappy-change areas require one soap dispenser above alternate paired sinks, one tissue dispenser on the wall above the changing mat, one paper towel dispenser on the wall adjacent to the sink and one hot air dryer on the wall adjacent to the sink.

— Electrical Services

Lighting

- Emergency lighting should be installed and specified in accordance with Local Authority guidelines.
- External lighting is required for entrance areas, playgrounds and car parks.
- Internal lighting should be warm and energy-saving rather than cold-looking, such as that provided by fluorescent fittings. Small spotlights, up-lighters and pendants can be used to highlight key areas and points of interest within the building such as noticeboards. Emergency lighting is required throughout and discussions with the fire brigade are recommended. Energy-saving fittings and bulbs must be provided.

This colourful image of a childcare centre in Reggio Emilia utilizes a range of different light fittings which enable variations and illumination which is high yet soft and user-friendly (design by ZPZ Architects).

Sockets

- It may seem obvious, but sockets should not be placed near water. None are required in the toilet/nappy-change area, although spurs for hot-air dryers will need to be fitted.
- In children's areas, electric sockets should be set at 1,200 mm (to underside) above the floor.
- Sockets in children's areas should be fitted with safety blanking-off plugs.
- Two double sockets are required per home base/activity area and in the office/head of centre's office. Sockets to other areas as required. Don't forget sockets for cleaners in the circulation areas.
- In the milk kitchen a double socket should be positioned above the worktop, away from the sink, and a single socket below the worktop for the fridge.
- Too many socket points are expensive and unnecessary.

Internet, computer, TV aerial and satellite dish

A TV aerial is usually required but may come as satellite or more usually a cable feed, depending on systems available. A DVD player in at least one of the activity areas and in the mini piazza is crucial.

Today (in the United Kingdom), internet broadband is readily available in most urban settings. It is an important requirement to aid staff–parent communications.

Emergency exits and equipment

Panic bars and fire escapes

- These must be on external doors if required by the fire officer. The age-old conflict between means of escape and security for the children must be considered. Panic pads or bars to external doors, *which can be opened by children unattended should be avoided*. An electro-magnetic system which can be adult-only operated is preferable.
- All external exit doors, if fire doors, should have either push-bar or thumb-turn access, sited at adult height. There is always some conflict between security in a nursery and ease of escape. We suggest that the client should be closely involved in any decisions.
- Fire detection, alarm and emergency lighting must be discussed with the fire officer.
- Please note that security must be carefully considered. An 1,800 mm high fence should screen all areas where the plot is adjacent to public paths or roads.
- The fence should not be climbable from inside or outside.
- Outside toy storage can also function as a play house. This may be a timber or brick construction, but care must be taken when designing the interior for free child access.

Finishes

- Toilet, nappy-change, kitchen and laundry areas should have appropriate wipe-clean surfaces. The Environmental Health Department should be consulted for local recommendations.
- Walls and ceilings should be decorated in vinyl matt emulsion paint.
- Woodwork should be decorated using egg shell paint. Dado rails should be fitted if possible to simplify redecorating.
- Stains can be used for wooden doors and cupboards in home bases.
- Non-slip flooring should be used in wet areas; hard floor elsewhere. In home bases rugs will be added.
- Sheet linoleum or vinyl is suitable for floors, typically 'Polysafe' or similar for general areas and a softer finish for home bases. The baby area could be a cushion floor.
- Carpet should be used in staffroom, reception and office areas.

Glazing

- The premises should receive natural sunlight during some part of the day. Attachments should be fitted for blinds, or curtain rails for each external window. All external windows must be openable and must have safety catches so they can be locked open at a safe width to avoid endangering passers-by – 100 mm or as agreed with the Local Authority.
- All windows should be double-glazed.
- Windows in child spaces should, where practical, be positioned low enough for children to see outside but high enough to accommodate radiators where underfloor heating is not specified. A suitable lower-edge height is 600–1,000 mm.
- Any glazing lower than 1 m above floor level must be toughened safety glass to the relevant BS standard, and must include some form of manifestation so children can see it easily.

Heating

- A heating system appropriate to the building is to be specified by the services engineer. Although underfloor heating can only really be fitted to new buildings rather than refurbishments, it is preferred. However, the system should be able to achieve 21°C in the home bases and 21°C in other areas within one hour when the external temperature is at −1°C. The heating system should have a seven-day digital programmer capable of running heating and hot water independently.
- Radiators should be low temperature or be covered with guards (small mesh) in all home bases and other child areas. Exposed pipework should be avoided or, if essential, protected. Children must not come into contact with surfaces at a temperature of more than 43°C.
- Underfloor heating is successful since children, especially babies, are playing at a low level.
- Hot water must be stored above 65°C. Delivery to children's rooms must be thermostatically mixed down to below 43°C.

A nursery brief

Milk kitchen

- The milk kitchen is ideally located adjacent to baby home bases. This may be of domestic quality and requires a worktop with an electric socket for a small under-worktop refrigerator and a kettle.
- Sinks are required for hand washing and a sink drainer for crockery, etc.
- There should be cupboards below and shelving above the worktop.
- Place a noticeboard on a cupboard door.

Noticeboards

- As many large 3 x 1 m noticeboards as possible in all areas except for the kitchen and staff toilets.
- Noticeboards may be of a fibre board type of material; one of the proprietary brands such as Sundeala is acceptable. Ideally they should be locally sourced.
- Noticeboards should be edged and should be able to take Blu-tack, staples and sellotape without damage.
- Positioning of noticeboards should be discussed with the head of centre.

Plumbing

- A macerator for nappy disposal requires a three-phase supply, specific drainage and water supply.
- A shower unit with a shower head adjustable down to 1 m above the base is useful. This should be installed in the children's toilet area associated with the baby or toddler home bases.
- Sinks in the toilet areas are normally specified as one per ten children, although Local Authorities have varied in their requirements. Low-level sinks (child level) with mirrors above are required. The sinks should be mounted at 550 mm to 600 mm above floor level. One soap dispenser to be fitted for every two sinks, above one of the pair of sinks. Social Services requirements may warrant toothbrush holders and toothbrushes – one for each child, but we would not recommend it.
- Child-height inset sinks or single sink drainers are required in each home base, with cupboards below and shelving above the sink. Soap dispenser and paper towels for hand drying. Worktops in all areas (except the baby home bases) should be at a height useable by children and not just adults, which should be 550 mm to 600 mm above ground level.
- A double sink and stainless steel double drainer is required.
- A separate ceramic hand-wash basin is required away from the main work area.
- A single sink and drainer is required in the milk kitchen.
- Ideally, a single stainless inset sink should be provided adjacent to the nappy-changing table, sited to the right of the nappy-change worktop with mixer taps away from the worktop.
- A sluice is not necessarily required. Although in the United Kingdom we use disposable nappies, for environmental reasons the decision may be made to change back to terry nappies via an American-style 'diaper

service'. Whatever the future decision, it is advisable to fit a sluice into the toilet or a nappy change area if practical.
- Taps to the children's areas should have chunky, easy-to-operate fittings. We prefer push-down taps so that children cannot leave the taps running.
- Water temperature to all plumbing where children have access to the 'hot' water supply must be thermostatically controlled so that the temperature does not exceed approximately 43°C. Water should not be stored at this temperature.
- Water to the hot supply in the kitchen, staffroom and laundry should be full temperature.

Adult toilets

- The toilet cubicles should be separated by melamine or laminated boarding, except where male and female are side by side, in which case there should be a solid (acoustically sound reducing) full-height partition. Individual toilet roll holders for each cubicle are required.
- The walls adjacent to the toilets should be tiled unless the toilet is boxed in with a wipe-clean melamine-surfaced sheet material. All toilets should be provided with a full-height door.
- One hot-air hand dryer in the staff toilet is required.
- A single staff toilet should be designed as a disabled unit, with fold-down baby change table.
- Doors should still have finger guards fitted or piano hinges.

Refuse areas

- The refuse area should be fenced off, and be located kitchen-side if possible, away from the main building entrance and children's play areas and able to accommodate two large bins or to accommodate whatever recycling system is in place locally.
- Larger nurseries require two bins for general refuse and one clinical waste bin, so the bin store area may need to accommodate three bins in total.

Security

Specification of the access controls and security systems must be discussed and agreed with a local supplier. Systems are evolving quickly and it is important to procure the best system on the market. Security is a key issue and must satisfy:

- allowing access to authorized personnel only;
- providing for ease of escape while preventing children letting themselves out of the building;
- security/fencing – if gates are incorporated in any part of the fencing, bolt closures must be fitted to the inside of the gate, top and bottom.

A nursery brief

Door bells

- The external door will be locked at all times except when a member of staff is on duty in the reception area.
- Access to the building at other times is by door bell. This should ring in the mini piazza, staffrooms and offices.
- When buildings are of more than one level, the door bell needs to ring out on all floors.
- The door and fire bells should not be so loud as to frighten the children.

Indoor safety

Worktops should have post-formed, rounded leading edges.

Outdoor safety

Outside gates should open outward, be securely hung and have two catches, one inside and one outside the gate. Fencing outside the play area must be of a safe design so that children are not able to climb over or under fencing.

Staffroom

- The size of the staffroom must be large enough for half of the staff to have a break in the room at any one time. This of course varies with the size of the nursery.
- Staff lockers or cupboards and coat hooks are necessary. A lockable cupboard should be fitted if lockers are not provided.
- A sink and drainer with electric sockets adjacent to the work surface and cupboards above and below is necessary.
- Electric sockets are necessary for training videos and television.
- One shelf above the sink is required.

Storage

In addition to built-in storage within the various home base areas, there is a great need for additional storage space. The laundry and kitchen have been specified. There needs to be a large (5–7 sq. metres) walk-in store for bulky items such as nappy boxes, paper and paint. This should be accessible from the 'play street' or the general circulation area.

Storage in the home bases

- A low-level sink and drainer mounted at 550 mm to 600 mm, with cupboard storage under and two built-in full-height cupboards with shelving in each home base is required as a minimum. Security locks/catches should be fitted to doors which children should not be able to open on their own.
- All low-level cupboards in home bases for babies and toddlers should be fitted with security catches. This includes nappy-change areas.

Telephones

- The siting of phone sockets should be discussed with the head of centre on an individual basis.
- At least two telephones should be installed: one in a central area, such as the mini piazza and another in the manager's office.
- Additional (not too loud) bells should be in strategic places.
- The system should be able to receive multiple lines and transfer calls to extensions.

The Learning Centre, up-state New York. The community room and main entrance beyond. Note the use of second-hand sofas and the open kitchen to provide a homely semi-domestic welcome.

CHECKLIST 5: KEY ASPECTS OF ARCHITECTURE FOR YOUNG CHILDREN

1. Is the architecture flexible and readily extendable? It will undoubtedly need to change and evolve over time as the users grow and change themselves during the decades of its life in use. Consider the building's affinity to the natural world, imagine how a flower grows and adapts to its surroundings.

2. Are the spaces interesting and engaging? The building should be designed as much from the child's perspective as from the adult's more practical viewpoint. Consider the scale of the child in comparison to the adult-sized world around them. As Dr Seuss Giegel pointed out, 'I know up on top you are seeing great sights, but down here at the bottom we, too, should have rights.'[5]

3. Is the architecture legible to its users? If children can see how the building is put together, the very act of discovery can elicit greater awareness of their surroundings. Consider the routes through the building as part of this knowledge rather than building separate rooms within which children are enclosed. Can a child find points within the building between which they can plan small, independent journeys?

4. Is there enough space? Guidelines set out in the National Standards[6] may not be adequate. Not having enough space can create tension and conflict, whereas fluid, open spaces encourage calm interactive play. Do the children have a range of semi-enclosed areas (possibly corners) that will support different activities and the need for some children to withdraw from the main group?

5. Are the outdoor spaces readily accessible? There should be run-in–run-out play with lots of different things to do outside. The garden can extend the field of learning, which is particularly valuable where the interior is cramped and confined. If there is no garden, even a balcony or terrace can have tremendous benefits extending the environment in use.

6. Do children and their parents feel safe and secure? Make the building feel welcoming yet secure from the outside world with clear separations between children's zones and the adult areas. When the centre has multiple users such as adult training or parenting classes, ensure that a workable yet fail-safe system of inner security is built into the overall planning strategy.

7. Acoustics: are play rooms and main children's activity rooms quiet, particularly an issue with single, open-plan play spaces. Consider acoustic treatment to ceilings, walls and floors.

8. Do the users understand the architecture? Effective consultation with the users should happen during the design and development of the new facility.

Notes

1 Environmental psychology: how to evaluate quality within the learning environment

1. Cervantes, Miguel De, *Don Quixote*, trans. Edith Grossman, Vintage, 2005, p. 503.
2. Fisher, Thomas, 'Architects behaving badly' 18 January 2005, blog article.
3. Frances E. Kuo is founder of the Human-Environment Research Laboratory at University of Illinois at Urban-Champaign.
4. Lynch, K., *The Image of the City*, MIT Press, 1960.
5. Stephen Kaplan is Professor of Psychology and Electrical Engineering and Computer Science at the Department of Psychology, University of Michigan.
6. Canter, D. and Woods, R., 'Judgements of people and their rooms'. *British Journal of Clinical Psychology* 13 (1974): 113–118.
7. Blaut, J. 'Environmental mapping in young children'. *Environment and Behaviour* 2(3) (1970): 335–349.
8. Emeritus Professor Christopher Spencer, Psychology Department, University of Sheffield, refers to research undertaken in 1989 which he does not value. Therefore refer to more recent publications, for example: *Environmental Psychology: Putting research into practice*, Cambridge Scholars Press, 2007. This refers to Advances in Child Behaviour, 25, pp. 157–199, 1989.
9. Moore, R., *Childhood's Domain*, Croom Helm, 1986, p. 36.
10. Ibid.
11. Ecological psychology, as seen in the work of Roger G. Barker, Herb Wright and associates at the University of Kansas during the 1950s. He was one of the first to recognize the importance of field studies rather than laboratory experiments of child behaviour. In his classic work *Environmental Psychology* (1968; out of print), he argued that human behaviour was radically situated – you couldn't make predictions about human behaviour unless you knew the context where it was played out.
12. Whitfield, Charles L., *Healing the Child Within*, Health Communications Inc, 1987, p. 1.
13. Russell, Bertrand, 'Education and social order' in *Bertrand Russell: The Spirit of Solitude*, Jonathan Cape, 1996.
14. Moore, G. 'The Children's *Physical* Environment Rating Scale (C*P*ERS): reliability and validity for assessing the physical environment of early childhood educational facilities'. *Children, Youth and Environments* 17 (2007): 35–56.

2 The sustainable nursery

1. Trent, David (ed.), *Reading Digital Culture*, Blackwell, 2001, p. 157.
2. Ward, C., *The Child in the City*, Bedford Press, 1990.
3. Letter to the *Daily Express*, 10 July 2002, p. 29.
4. This refers to the case which as I write is dominating the news agenda in December 2008: that of Baby P (he cannot be named for legal reasons). He suffered more than 50 injuries at the hands of his mother, her boyfriend and a lodger. There was no-one around on a daily basis to

look after him. If Baby P had been placed in full-time nursery from the age of three months his death would almost certainly have been avoided. There are currently 450 children on the at-risk register in his Local Authority, Haringey. Sadly, it is a case of too few resources to protect at-risk children – a civilized country should do more.

5 Clarke-Stewart, A., 'Infant daycare: malignant or maligned'. *American Psychologist* 44 (1989): 266–273.
6 Schweinhart, L.J., Montie, J., Xiang, Z., Barnett, W.S., Belfield, C.R. and Nores, M., *Lifetime Effects: The HighScope Perry Preschool Study Through Age 40*, HighScope Press, 2005.
7 Kline, Stephen, *Out of the Garden: Toys and Children's Culture in the Age of TV Marketing*, verso, 1993, p. vii
8 Holt, N. HighScope UK, *The HighScope Approach in Practice*, Routledge, 2010.
9 Boubekri, Mohamed 'Lighting design' in M. Dudek (ed.), *Schools and Kindergartens: A Design Manual*, Birkhauser, 2008, p. 34.
10 From a document published by Fulcrum Consulting, London, entitled *Schools: Sustainable Design*, 2004.
11 From the architect's statement, sent to the author.
12 Bettelheim, Bruno, quoted by Annalia Galdini at conference on 'The Organisation of Space in Services for Children', Madrid, 27–28 November 1992.
13 Speech given at the American University, Washington, DC, 10 June 1963.

3 The natural child

1 Letter from Baroness Greenfield, Director of the Royal Institution, and 270 others, February 2008, *Daily Telegraph*.
2 Aric Sigman quoted in Hall, Sarah, 'Watching television harms toddlers, says psychologists'. *Guardian*, 24 April 2007, p. 5.
3 From *7Cs*, an informational guide to young children's outdoor play spaces. Available online: http://westcoast.ca/playspaces/outsidecriteria/7Cs.html.
4 Wood, D., 'Ground to stand on: some notes on kids' dirt play'. *Children's Environments* 10 (1) (1993): 16.
5 For an account of Alison Clarke's research, refer to Dudek, Mark 'Talking and listening to children', *Children's Spaces*, Architectural Press, 2005, pp. 1–13.
6 Christopher Day quoted in Dudek, Mark, *Kindergarten Architecture*, Spon, 2000, p. 28.
7 Heft, H., 'Affordances of children's environments: a functional approach to environmental description'. *Children's Environment Quarterly* 5 (3) (1988): 29–37.
8 Sheridan, Mary D., *Play in Early Childhood*, Routledge, 1977, reprinted 1999, p. 4.
9 Sheridan describes this as 'informal learning', which suggests that it is relaxed and indeterminate; however, for most children research is the most intense and sustained form of play, and therefore the most valuable. Children should be given the space, resources and time to engage. From Sheridan, *Play in Early Childhood*, 1999.
10 Sheridan, *Play in Early Childhood*, p. 16.
11 Tuan, Yi-fu, 'Language and the making of place: a narrative-descriptive approach'. *Annals of the Association of American Geographers* 81 (4) (1991): 685–695.
12 Cobb's book explores the idea that genius is shaped by the imagination of the child. It contains a collection of autobiographies and biographies of creative people, as well as her observations of children's play. She sees the child to be innately connected with the natural world. Edith Cobb, *The Ecology of Imagination in Childhood*, Spring Publications, 1993.
13 Herrington, Susan and Studtmann, Ken, 'Landscape interventions: new directions for the design of children's outdoor play environments', *Landscape and Urban Planning* 42 (1998): 191–205.
14 Nicholson, S., 'How not to cheat on children: the theory of loose parts'. *Landscape Architecture* 62(1) (1971): 30–34.
15 Taken from *7Cs*, an informational guide to young children's outdoor play spaces. Available online: http://westcoast.ca/playspaces/outsidecriteria/7Cs.html.
16 Sigman, A., 'The Biological Effects of Daycare', *The Biologist*, 58(3), pp. 28–29.
17 Bilton, Helen, *Outdoor Play in the Early Years*, David Fulton, 2002, p. 28.

4 A historical overview

1 Eisenman, Peter, 'Post-Functionalism' in Nesbitt, Kate (ed.) *Theorizing A New Agenda for Architecture: An Anthology of Architectural Theory, 1965–1995*, Princeton Architectural Press, 1996, p. 78.
2 Markus, Thomas A., *Buildings and Power*, Routledge, 1993, p. 78.
3 Pestalozzi, J.H., *Leonard and Gertrude: A popular story written originally in German; translated into French and now in English; with the hope of being useful to ALL CLASSES OF SOCIETY*, now available online.
4 Ashwin, C., *Drawing and Education in German-speaking Europe 1800–1900*, dissertation, The Institute of Education, London University, 1980, p. 46.
5 Grimm, J., and Grimm, W., *Grimm's Fairytales illustrated by Arthur Rachham*, first published 18R, reprinted by The Folio Society, 2011.
6 Kaufmann, E., 'Form became feeling: a new view of Froebel and Wright', *Journal of the Society of Architectural Historians*, 40, (1981): 130–137.
7 Stewert, W.A.C. and McCann, W.P. *The Educational Innovators, 1750–1880*, St Martin's Press, reprint 1970, p. 67.
8 McMillan, Margaret, *The Camp School*, Allen & Unwin, 1919, p. 28.
9 Saint, A., *Towards a Social Architecture: The Role of School Architecture in Post-war England*, Yale University Press, 1987, p. 12.
10 McMillan, Margaret, *The Camp School*, Allen & Unwin, 1919, p. 30.
11 Isaacs, Susan, *Intellectual Growth in Young Children*, Routledge, 1930, p. 35.
12 Monk, Ray, *Bertrand Russell: The Spirit of Solitude*, Jonathan Cape, 1996, p. 220.
13 Swiss developmental psychologist Jean Piaget (1896–1980) related the intellectual development of the infant to the gradual discovery of how the environment can be altered. This comes about through imitation and play. For several years he observed his own children very closely, and was the first to suggest that the new-born baby's world is a confusion of meaningless images. At the age of four months, the baby becomes aware of an outside world distinct from his or her own body. However, what is missing is an awareness that when something disappears from view, it still exists; to an infant, anything not visible is non-existent. As children develop, they learn to retain an impression of things that have gone from view, and are then ready for a more elaborate form of play. Intelligence develops in a series of stages; as each stage unfolds the child must keep-up with the earlier level of mental understanding to reconstruct it towards a higher level of understanding.
14 Book review by Julian Grenier in *Nursery World*, 3 December 2009.
15 van der Eyleen, W. and Turner, B. *Adventures in Education*, Penguin, 1969, p. 55.
16 Montessori, M., *The Montessori Method: Scientific Pedagogy as Applied to Child Education in "The Children's Houses" with Additions and Revisions by the Author*, second edition, Frederick A. Stokes Company, MCMXII, 1912.
17 Rita Kramer, *Maria Montessori: A Biography*, Da Copo Press, 1988.
18 Corigan, C., OECD thematic review of early childhood education and care background report, 2002, record no. 109.
19 Wright, F.L., *Frank Lloyd Wright: An Autobiography*, Duel Sloan and Pearce, 1943, pp. 13–14.
20 Wright, F.L. in *The Cause of Architecture: VI*, p. 200, quoted in Sloan, Julie L., *Light Screens: The Leaded Glass of Frank Lloyd Wright*, Rizzoli International Publications Inc., 2001.
21 Giulio Ceppi and Michele Zini (eds), *Children, Spaces, Relations: Metaproject for an Environment for Young Children*, Reggio Children/Domus Academy, 1998.
22 Penn, Helen, *Discovering Nurseries*, Paul Chapman Publishing Ltd, 1997.

5 A nursery brief

1 Robin Moore is based at the North Carolina State University, where he is Director of Natural Learning Initiative, Professor of Landscape Architecture, College of Design.
2 HighScope is a tailored early-years curriculum form, as dicussed in Chapter 2.
3 Roger Ulrich is a pioneer of evidence-based design and Professor of Architecture at Texas A&M University.

Notes

4   Refer to Building Bulletin 93, *Acoustic Design in Schools: A Design Guide*, The Stationary Office. Although it does not deal with early-years environments specifically, its advice for classrooms in use is adaptable.
5   *Dr Seuss: A Classic Treasury*, Harper Collins Publishers, 2006, p. 56.
6   Sure Start – Full day care, National Standards for under 8, day care and childminding, Crown copyright 2003. Produced by the Department for Education and Skills and Department for Work and Pensions.

Index

Page numbers in **bold** refer to illustrations.

Activities 39; *see also* Children's activity areas
Acoustic design 74, 190
ADHD and ADD 18
Adjacency diagrams 160
Affordances 21, 22
Architects behaving badly 11
Arkitema 161
Arroyo, E. 135
Avery Coonley Playhouse 119; *see also* Lloyd Wright, F.

Baby daycare 159
Barker, R. 22
Beaudouin and Lods 129
Behnisch and Partners 29, 133
Berlin 47; *see also* Kindergarten Jerusalemur Strasse
Bettelheim, B. 76
Big society 37
Bilbao Sondika Kindergarten 135
Blaut, J. 21
Bolles Wilson 133, **140–141**; *see also* Frankfurt
Bowlby, J. 36, 124
Break the will of the child 116
Building Regulations, UK 68
Burke, C. 5
Burley Children's Centre, Leeds 27, **65–68**, 73

CABE 11
Canter and Woods 20
Cape Cod 33
Catalan 135
Child-centred learning 110
Child in the city 32
Child-oriented 120
Children Act 159
Children's activity areas 162
Children's centres 53
Chollet and Mathan 130
Churchill, W. 15

Clark, A. 5, 84
Clarke Hall, D. 139
Cloakrooms 172
Cobb, E. 96
Colour form language 141
Columbia 50; *see also* El Porvenir Kindergarten, Bogota 50
Complex scene 19
Cotrell and Vermeulen Architecture 131
Crowley, M. 138
Curriculum 93; *see also* activities and HighScope

Daily Telegraph 79
Darvezeh, Z. 21
Day, C. 84, 96
Denmark 35, 105
Designated zones 172
Devlin, C. 63
Dewey, J. 129
Digital playground 79
Disadvantaged home environments 39
Discovering Kids Playgroup, Loup 14, 63–65
Dr Seuss 190
Domus Academy Research Centre 42, 137
Dovedale 102
Drygate, Glasgow 113

Ecoles Maternelle Ville De Cachan 130
Ecological psychology 22, 23
Ehn, K. 127
Eisenman, P. 112
El Porvenir Kindergarten, Bogota 50
English Heritage 11
Environmental psychology 14, 18, 22
Erect Architecture **144–147**; *see also* London
EU Peace and Reconciliation Fund 63
European Community 151
Every Child Counts 46

Index

Familio Children's Centre, Mansfield 53–57
Ferrari 134
Festival of Britain 139
Fisher, T. 11
Fondation de France Ministere Des Affairs Sociales 88
Form becomes feeling 119
Frankfurt 7, 115, **140–141**; Bolles Wilson 6; Frankfurt Kindertagesstatte 140–141; Kindertagesstatte 7; Pestalozzi, J. 115; Toyo Ito 6
Froebel, F. 114–118, 120, 126, 135; Avery Coonley Playhouse 119; Froebel Company 116; Froebel Gifts 116, 118, 119; Lloyd Wright, F. 116; Tavistock Square 116; Teachers 118; Yverdun 133
Frey, K. 62, 78
Fry M., Atkinson R., James C. H. and Wornum G. 131–132
Fulcrum 76

Garden of Eden 114, 115, 126
Glasgow Infant School 114
Goldfinger, E. 130, 138, 139
Grantham Children's Centre 157
Green is good for you 16–17
Grey water concept 75
Grimm's Fairytales 115
Gropius, W. 128
Gross motor skills 93

Harry Potter 93
Heft, H. 87
Herrington, S. 5, 96, 98, 100
HighScope 2, **40–45**, 53
Hillingdon Children's Centres 5
Home base 162, 188
Hospital buildings 18
Howe and Lescaze 128

Industrial Revolution 112
Isaacs, S. 123
Itard, J and Seguin, E. 122
Ito, T. 133

Jordanhill College of Education, Glasgow 113
Jung, C. 24
Jurgow Park Nursery **153–156**

Kansas 22
Kaplan, S. 19
Kennedy, J. F. 77
Kindergarten Jerusalemur Strasse 47
Kitchens 89, 158, 168, 177
Kline, S. 41
Kuo, F. 18

Labyrinth 104
Labour government 36
Land Rover Discovery 100
Le Corbusier 106, 128, 139
Legibility 20

Lerner, R. 24
Leuzinger, H. 132
Linguistic skills 93
Lloyd Wright, F. 116–119, 133–134; Avery Coonley Playhouse 119, 133; Centennial in Philadelphia 134; Form becomes feeling 119; Froebel gifts 133; Prairie House 134
London: Cherry Lane Children's Centre, Hillingdon 19, 30, 38, 176; Cornerstone Centre at Yiewsley 25; Effra Early Years Centre, Brixton 60, 62–63, 73; Festival of Britain 139; Kilburn Grange Park Adventure Playground **144–147**; Lavender Children's Centre 68; Little Cherubs Kindergarten 47; Nursery at Kensal Rise 131; Paradise Park Children's Centre 71–73; Portman Centre, Westminster 15, 105, 106, **164–165**; Princess Diana Memorial Fountain 17; Stanley Infant and Nursery School 70, 152, 169; Tavistock Square 116; Windham Nursery, Richmond 8–9, 59–60; Zoo 43
Long, R. 5
Loxos baby change unit 167
Lustenau Kindergarten 62–63, 78
Lynch, K. 19, 20
Lurcat, A. 129

Malaguzzi, L. 134, 137
Maltinghouse School 123
Marseille Unité 128
Mazzanti, G. 50
McMillan, M. 120–123; Lewisham Local Studies 121; McMillan Open Air Nursery School, Brighton 121; Nursery School Association 123; Nursery School Movement 122; Open-air School, Deptford 121
MDF 112
Meadowcroft Griffin Architects 58
Meadowcroft, P. 59
Mies Van Der Rohe 139
Memories of childhood 23
Miller, A. 24
Modern Movement 130
Montessori, M. 125–127,129; Casa dei Bambini, Rome 126; Haus der Kinder, Vienna 125; Karl Marx Hof, Vienna 127; Montessori Method, 125; Phases of development 125
Moore, R. 22, 28–29, 148

Nant-Y-Cwm Steiner Nursery 84–86, 96; see also Steiner
NCH 79
New Lanark 117; see also Owen, R.
Nicholson, S. 103

Observation room 163
OECD 133
Open-air nursery schools 122
Orphanage 34

Outdoor classroom 110
Owen, R. 117, 120; New Institution for the Formation of Character, 117; Naturalistic objects 118; *see also* New Lanark

Pacific Oaks 34
Penn, H. 135
Pestallozi J. H. 114, 120; Burgdorf 114; Christoph Buss 114; Froebel, F. 115
Peter Pan 101
Pansa, S. 10
Piaget, J. 44, 123–124
Play sessions and events 39
Play street 175
Playtraces 22
Play yard 51
Poetic language 97
Portakabin 77
Pre-school education 39
President Johnson, L. 40

Quale, M. 65

Raleigh, N Carolina 82–83, 96
Reggio Emilia, 41, 132, 134, **136–137**, 170; Scuola dell'Infanzia San Felice, Modena 134, 136–137; Scuola dell'Infanzia Diane 42; Malaguzzi, L. 134, 137; Mealtimes 170; ZPZ Partners 134
Ritalin 113
Royal Commission 62
Romanticism 114
Run-in–run-out play 109, 157
Rural Development Council, N Ireland 65
Ruskin 15
Russell, B. 26

Sand-pit 150, 173
Stairs 151
St Pancras Station 40
Scandinavian 35, 36, 40
Screenager 31
Second skin 43
Second World War 112
Security 30
Sensory garden 109
Sensory stimulants 107
Sheridan, M. D. 89–95
Siamang 13, 22
Sigman, A. 80–81, 107
Sinking Boat Kindergarten 29

Social pedagogy 35
Spencer, C. 11, 23–24, 30,
Staffroom 69
Steiner, R. 135, 137; Spiritual metamorphosis 135; Sustainable future 137
Stow, D. 113–114,
Studtmann, K. 96
Sure Start 41, 68, 73
Sustainable building 69
Synomorphy 20, 23
Synthetic 112
Sweden 35
Symbolic 93

Tactile experiences 98
Tatra Mountains 143
Television 81
Thatcher, M. 36, 133
Thomas Coram Research Unit 5
Tolladine Surestart and Community Centre 57–58
Toy store 180
Tuan, Y. 96

Ulrich, R. 18
United States 52; Avery Coonley Playhouse, LA 119; Caryl Peabody Nursery School 128; Corning Centre, Corning, NY 189; Elementary school classrooms 52; Oak Lane Country Day School, Pennsylvania 128; Wiltheis Kindergarten School. Seattle 118
Units of profit 111
University of British Columbia 5, 100
University of Iowa 99
University of Sheffield 5
Urridaholtskoli, Iceland 161

Vancouver 100
Viborg, Denmark 143
Visual transparency 149

Ward, C. 32
Water play 173, 17
Wayfinding 17
Whitfield, C. L. 24
Wiltheis, C and T. 118
Winnicot, D. 24
Wood, D. 82, 96, 98
Workshop environment 7

ZPZ Partners 134